# The Manufacture of Mineral and Lake Pigments

You are holding a reproduction of an original work that is in the public domain in the United States of America, and possibly other countries. You may freely copy and distribute this work as no entity (individual or corporate) has a copyright on the body of the work. This book may contain prior copyright references, and library stamps (as most of these works were scanned from library copies). These have been scanned and retained as part of the historical artifact.

This book may have occasional imperfections such as missing or blurred pages, poor pictures, errant marks, etc. that were either part of the original artifact, or were introduced by the scanning process. We believe this work is culturally important, and despite the imperfections, have elected to bring it back into print as part of our continuing commitment to the preservation of printed works worldwide. We appreciate your understanding of the imperfections in the preservation process, and hope you enjoy this valuable book.

THE MANUFACTURE OF
MINERAL AND LAKE PIGMENTS

CONTAINING DIRECTIONS FOR THE MANUFACTURE OF ALL
ARTIFICIAL ARTISTS' AND PAINTERS' COLOURS,
ENAMEL COLOURS, SOOT AND METALLIC PIGMENTS

A Text-Book for
Manufacturers, Merchants, Artists and Painters

BY

Dr. JOSEF BERSCH

TRANSLATED FROM THE SECOND REVISED EDITION
BY
ARTHUR C. WRIGHT, M.A. (Oxon.), B.Sc. (Lond.)
FORMERLY ASSISTANT LECTURER AND DEMONSTRATOR IN CHEMISTRY AT THE YORKSHIRE
COLLEGE, LEEDS; AUTHOR OF "SIMPLE METHODS FOR TESTING PAINTERS'
MATERIALS"

WITH FORTY-THREE ILLUSTRATIONS IN THE TEXT

LONDON
SCOTT, GREENWOOD & SON
"THE OIL AND COLOUR TRADES JOURNAL" OFFICES
8 BROADWAY, LUDGATE HILL, E.C.

CANADA: THE COPP CLARK CO., LTD., TORONTO
UNITED STATES: D. VAN NOSTRAND CO., NEW YORK
1901
[The sole right of publishing this work in English rests with Scott, Greenwood & Son]

PREFACE.

WE know hardly another branch of chemical technology which has made such remarkable advances of late as the manufacture of colours; a large number of pigments have been recently discovered, distinguished by beauty of shade and permanence. Chemists are continually endeavouring to replace handsome and poisonous colours by others equally handsome but non-poisonous.

In writing this work I have endeavoured to give it such a character that it may be a text-book for the practical man, only those methods have been given which certainly lead to a good result; in the case of new pigments I have only described methods of preparation which I have myself found to give good results.

Since it cannot be imagined that any one quite ignorant of chemistry could successfully manufacture colours (colours being always made by chemical processes which occasionally are rather complicated), I have, therefore, presupposed a knowledge of the principles of chemistry. In the short sketch of the chemical properties of the raw materials used in making ordinary pigments, the principal properties of the materials used by the colour-maker are given.

In order to make this treatise useful to dealers in and consumers of colours, the chapters dealing with the examination of pigments have been so arranged that the nature or adulteration of a pigment can be determined quickly and with certainty by any one.

Recipes, which originated at a time when empiricism ruled in chemistry, have been omitted, since they would only detract from the clearness of the matter.

As far as it is possible I have avoided the "recipe fetish," and have endeavoured to make clear to the reader the chemical processes to which regard must be had in the manufacture of the different pigments. Since the appearance of the first edition there have been many valuable innovations in the mineral colour industry, to which regard has been given in preparing this second edition in so far as they possess a really practical value.

A critical examination of proposals and formulæ, which are found in large numbers in the journals, has been avoided, since I wished to keep for my book that character of a reliable text-book and book of reference which was ascribed to it in the form of its first edition.

Dr. JOSEF BERSCH.

TRANSLATOR'S PREFACE.

IF excuse be needed for presenting a translation of Dr. Bersch's book at so long an interval after the publication of the original (1893), it must be sought in the paucity of the English literature on the subject. It is hoped that the practical nature of the work will make it acceptable to the English reader.

The subject-matter of the original has been preserved in the translation without alteration or addition, with the exception of an unimportant change in the order of arrangement.

The metric system of weights and measures has been used throughout; for the convenience of those who are not familiar with this system, directions are given in an appendix for converting into English weights and measures.

The section on paint grinding (Chapter LXIX.) is perhaps somewhat incomplete; for a more detailed and modern account of this branch of the subject the reader is referred to *Practical Paint Grinding*, by Mr. J. Cruickshank Smith, B.Sc., shortly to be issued by the same publishers.

A. C. WRIGHT.

HULL, *January*, 1901.

TABLE OF CONTENTS.

	PAGE
CHAPTER I. INTRODUCTION	1
CHAPTER II. THE PHYSICO-CHEMICAL BEHAVIOUR OF PIGMENTS	8
CHAPTER III. RAW MATERIALS EMPLOYED IN THE MANUFACTURE OF PIGMENTS	14
CHAPTER IV. ASSISTANT MATERIALS	16

    Water—Chlorine—Ammonia—The Hydrometer—Sal Ammoniac or Ammonium Chloride—Ammonium Sulphide.
    Acids.—Hydrochloric Acid—Sulphuretted Hydrogen—Sulphuric Acid: Oil of Vitriol, Nordhausen Sulphuric Acid—Nitric Acid—*Aqua Regia*—Carbon—Carbonic Acid Gas.
    Organic Acids.—Acetic Acid—Oxalic Acid—Tartaric Acid.

CHAPTER V. METALLIC COMPOUNDS 32

    Alkalis.—Potassium Compounds—Potassium Carbonate—Potassium Hydroxide—Potassium Nitrate—Potassium Bitartrate—Potassium Bichromate—Potassium Sodium Chromate—Chrome Alum—Potassium Ferrocyanide—Potassium Ferricyanide—Sodium Salts—Sodium Carbonate—Sodium Hydroxide—Sodium Thiosulphate—Sodium Chloride—Salts of the Alkaline Earth Metals.
    Calcium Compounds.—Calcium Oxide—Calcium Hydroxide—Calcium Carbonate—Calcium Sulphate—Calcium Phosphate—Magnesium Carbonate—Barium Compounds—Barium Chloride.
    Aluminium Compounds—Aluminium Sulphate.
    The Alums.—Potassium Aluminium Alum—Roman Alum—Soda Alum—Ammonia Alum—Alumina and Hydrate of Alumina.
    Compounds of the Heavy Metals.—Zinc Compounds—Cadmium Compounds—Iron Compounds—Ferrous Sulphate—Ferrous Chloride—Manganese Compounds—Nickel Compounds—Cobalt Compounds—Chromium Compounds—Molybdenum, Tungsten and Vanadium Compounds—Antimony Compounds—Bismuth Compounds—Tin Compounds—Arsenic Compounds—Lead Compounds—Lead Sulphate—Lead Nitrate—Lead Acetate—Basic Lead

TABLE OF CONTENTS.

PAGE

Acetate—Lead Chloride—Copper Compounds—Copper Sulphate—Copper Nitrate—Copper Acetate—Mercury Compounds—Mercurous Nitrate — Mercuric Nitrate — Mercurous Chloride — Mercuric Chloride—Silver Compounds—Gold Compounds.

CHAPTER VI. THE MANUFACTURE OF MINERAL PIGMENTS . . . 71
White Mineral Pigments—White Lead.

CHAPTER VII. THE MANUFACTURE OF WHITE LEAD 77
Manufacture of White Lead from Metallic Lead.—Dutch Process: Casting the Lead into Sheets—Building up the Stacks—Removal and Grinding of the White Lead—White Lead Mills—Hard White Lead—Soft White Lead.—German Process.—French Process: Preparation of the Solution of Basic Lead Acetate—Preparation of the Carbonic Acid and Precipitation of the White Lead—Manufacture of White Lead by Means of Natural Carbonic Acid.—English Process.—Other Methods.—Oxychloride White Lead: Lead Sulphite—Lewis & Bartlett's White Lead Pigment.—White Lead-Antimony Pigments: Lead Antimonite—Lead Antimonate.

CHAPTER VIII. ENAMEL WHITE. 116
Lithopone.

CHAPTER IX. WASHING APPARATUS 120
Filter Presses.

CHAPTER X. ZINC WHITE 126
Griffith's Zinc White—Tungsten White.—White Antimony Pigments: Antimony Trioxide—Antimony Oxychloride—Bismuth White—Tin White—Manganese White—Magnesia White or Mineral White—Annaline.

CHAPTER XI. YELLOW MINERAL PIGMENTS. 133

CHAPTER XII. CHROME YELLOWS 134
Lead Chrome Yellow—Preparation of the Lead Solution—Precipitation of the Chrome Yellow—The Pale Chrome Yellows.

CHAPTER XIII. LEAD OXIDE PIGMENTS 143
Red Lead, Minium.

CHAPTER XIV. OTHER YELLOW PIGMENTS . . . .
Cassel Yellow—Montpellier Yellow—Turner's Yellow or English Yellow—Naples Yellow—Antimony Yellow—Calcium Chrome Yellow—Barium Yellow, Yellow Ultramarine or Permanent Yellow—Zinc Chrome Yellow—Cadmium Chrome Yellow—Cadmium

TABLE OF CONTENTS.

PAGE

Yellow—Lead Iodide—Mars Yellow—Siderin Yellow—Aureolin—Tungsten Yellow—Nickel Yellow—Mercury Yellow or Turpeth Mineral—Yellow Arsenic Pigments—Lead Arsenite—Thallium Pigments.

CHAPTER XV. MOSAIC GOLD 160
Chrysean.

CHAPTER XVI. RED MINERAL PIGMENTS 163
Vermilion—Black Mercuric Sulphide—Red Mercuric Sulphide.

CHAPTER XVII. THE MANUFACTURE OF VERMILION 166
Dry Method: Chinese Vermilion.—Wet Method: Firmenich's Method—Liebig's Process—Infusible White Precipitate—Electrolytic Process—Mercuric Iodide.

CHAPTER XVIII. ANTIMONY VERMILION 178
Appendix—Antimony Blue.

CHAPTER XIX. FERRIC OXIDE PIGMENTS 180
Vogel's Iron Red—Macay's English Red—Indian Red.

CHAPTER XX. OTHER RED MINERAL PIGMENTS 186
Chrome Red or Chrome Vermilion—Cobalt Red—Cobalt Magnesia Red—Cobalt Arsenate—Chromium Stannate—Silver Chromate.

CHAPTER XXI. PURPLE OF CASSIUS 190
Magnesia Gold Purple—Alumina Gold Purple.

CHAPTER XXII. BLUE MINERAL PIGMENTS—IRON-CYANOGEN PIGMENTS . 194
Chinese Blue — Prussian Blue — Mineral Blue — Soluble Prussian Blue—Special Processes for the Manufacture of Chinese Blue—Turnbull's Blue—Antwerp Blue.

CHAPTER XXIII. ULTRAMARINE 204

CHAPTER XXIV. THE MANUFACTURE OF ULTRAMARINE 211
Preparation of Mixtures for Ultramarine — Ultramarine Violet—Chlorine and Steam Process—Hydrochloric Acid and Air Process—Ammonium Chloride Process—Pale Blue Ultramarine—Ultramarine Red.

CHAPTER XXV. BLUE COPPER PIGMENTS 226
Bremen Blue and Green—Neuberg Blue—Lime Blue—Payen's Mountain Blue—Oil Blue—Copper Hydroxide.

CHAPTER XXVI. BLUE COBALT PIGMENTS 230
Cobalt Blue, Thénard's Blue, Cobalt Ultramarine, King's Blue, Leyden Blue—Cæruleum—Cobalt Zinc Phosphate.

	PAGE
CHAPTER XXVII. SMALTS	233

Preparation of the Charge—Fusion of the Charge—Grinding the Fused Mass—Tungsten Blue—Tessié du Motay's Blue—Molybdenum Blue.

CHAPTER XXVIII. GREEN MINERAL PIGMENTS — GREEN COPPER PIGMENTS 240
  Green Copper Pigments: Copper Carbonate — Copper Arsenite— Scheele's Green — Swedish Green — Brunswick Green — Green Verditer—Neuwied Green—Copper Oxychloride.

CHAPTER XXIX. EMERALD GREEN 244
  Manufacture of Emerald Green from Verdigris—Manufacture of Emerald Green from Copper Sulphate—Mitis Green or Vienna Green—Copper Stannate—Kuhlmann's Green—Elsner's Green— Casselmann's Green—Lime Green—Patent Green—Copper Borates—Copper Silicate (Egyptian Blue).

CHAPTER XXX. VERDIGRIS 252
  Blue Verdigris—Distilled or Crystallised Verdigris—German Verdigris.

CHAPTER XXXI. CHROMIUM OXIDE 260
  Chrome Green.

CHAPTER XXXII. OTHER GREEN CHROMIUM PIGMENTS . . . 264
  Guignet's Green—Emerald Green—Chrome Green Lake—Turkish Green—Leaf Green.—Chromium Phosphate Pigments: Arnaudan's Green— Plessy's Green — Schnitzer's Green — Chromaventurine —Chrome Blue (Garnier).

CHAPTER XXXIII. GREEN COBALT PIGMENTS 268
  Cobalt Green.

CHAPTER XXXIV. GREEN MANGANESE PIGMENTS 270
  Manganese Green—Rosenstiehl's Green—Böttger's Barium Green— Manganous Oxide—Manganese Blue.

CHAPTER XXXV. COMPOUNDED GREEN PIGMENTS 273
  Chrome Green—Elsner's Chrome Green—Silk Green—Natural Green— Non-arsenical Green.

CHAPTER XXXVI. VIOLET MINERAL PIGMENTS 276
  Chrome Chloride—Manganese Violet—Tin Violet, Mineral Lake— Copper Violet, Guyard's Violet.

CHAPTER XXXVII. BROWN MINERAL PIGMENTS 279
  Lead Brown—Manganese Brown—Pyrolusite Brown—Prussian Brown Iron Brown—Copper Brown—Chrome Brown—Cobalt Brown.

TABLE OF CONTENTS.

PAGE

CHAPTER XXXVIII. BROWN DECOMPOSITION PRODUCTS 283
Humins—Bistre.

CHAPTER XXXIX. BLACK PIGMENTS 285
Charcoal Blacks: True Charcoal Black—Vine Black—Vine Black from Wine Lees—Vine Black from Pressed Grapes—Bone Black or Ivory Black.

CHAPTER XL. MANUFACTURE OF SOOT PIGMENTS 294
The Manufacture of Soot Blacks on the Large Scale.

CHAPTER XLI. MANUFACTURE OF LAMP BLACK 307
Calcination of the Soot—Pine Black.

CHAPTER XLII. THE MANUFACTURE OF SOOT BLACK WITHOUT CHAMBERS. 313

CHAPTER XLIII. INDIAN INK 316
Neutral Tint Black—Appendix: Black Mineral Pigments—Chrome Copper Black—Chrome Black.

CHAPTER XLIV. ENAMEL COLOURS 319
White Enamels—Coloured Enamels: Yellow Enamels—Red Enamel—Blue Enamels—Green Enamel—Violet Enamel—Black Enamel.

CHAPTER XLV. METALLIC PIGMENTS 326
Shell-Gold—Shell-Silver—Imitation Silver.

CHAPTER XLVI. BRONZE PIGMENTS 329
Electrolytic Copper Bronze—Tungsten Bronze Pigments.

CHAPTER XLVII. VEGETABLE BRONZE PIGMENTS 339
Appendix: The Brocade Pigments.

PIGMENTS OF ORGANIC ORIGIN.

CHAPTER XLVIII. LAKES 343

CHAPTER XLIX. YELLOW LAKES 348
Dutch Pink—Weld Lake—Gamboge Lake—Prepared Gamboge—Fustic Lake—Quercitron Lake—Purree or Indian Yellow—The Colouring Matter of Saffron—Colouring Matter of Gardinia Grandiflora.

CHAPTER L. RED LAKES 354
Cochineal and Carmine

CHAPTER LI. THE MANUFACTURE OF CARMINE 357
Cenette's Method—Munich, Vienna, Paris or Florentine Lake—Ammonia-Cochineal.

TABLE OF CONTENTS.

	PAGE
CHAPTER LII. THE COLOURING MATTER OF LAC	363

Lac Dye.

CHAPTER LIII. SAFFLOWER OR CARTHAMINE RED 366
Safflower Carmine—Alkanet.

CHAPTER LIV. MADDER AND ITS COLOURING MATTERS 370
Garancin—Garanceux—Madder Extract—The Constituents of Madder.

CHAPTER LV. MADDER LAKES 375
Madder Carmine.

CHAPTER LVI. MANJIT (INDIAN MADDER) 378
Chica Red, Curucuru, Carajuru—Bigonia Chica.

CHAPTER LVII. LICHEN COLOURING MATTERS 380
Archil—French Purple—Cudbear—Litmus.

CHAPTER LVIII. RED WOOD LAKES 384

CHAPTER LIX. THE COLOURING MATTERS OF SANDALWOOD AND OTHER
DYE-WOODS 388

CHAPTER LX. BLUE LAKES 390
Indigo—The Constituents of Indigo.

CHAPTER LXI. INDIGO CARMINE 394
Indigo Mills—Blue Lake.

CHAPTER LXII. THE COLOURING MATTER OF LOGWOOD 398
Logwood Extract—Kohlrausch's Process for Obtaining Concentrated
Extracts of Colouring Matters and Tannins.

CHAPTER LXIII. GREEN LAKES 409
Chlorophyll—Sap Green—Chinese Green, Lokao—Charvin's Green.

CHAPTER LXIV. BROWN ORGANIC PIGMENTS 414
Asphaltum—Sepia.

CHAPTER LXV. SAP COLOURS 416

CHAPTER LXVI. WATER COLOURS 419
Moist Water Colours.

CHAPTER LXVII. CRAYONS 423
Crayons for Earthenware.

CHAPTER LXVIII. CONFECTIONERY COLOURS . . . . .

CHAPTER LXIX. THE PREPARATION OF PIGMENTS FOR PAINTING .
Paint Mills.

TABLE OF CONTENTS. xv

PAGE

CHAPTER LXX. THE EXAMINATION OF PIGMENTS 434
 Mineral Pigments—Examination with the Blowpipe—Reactions of the White Pigments—Reactions of the Yellow Pigments—Reactions of the Red Pigments—Reactions of the Blue Pigments—Reactions of the Green Pigments—Reactions of the Brown Pigments—Reactions of the Black Pigments.

CHAPTER LXXI. EXAMINATION OF LAKES 445
 Reactions of the Organic Colouring Matters.

CHAPTER LXXII. THE TESTING OF DYE-WOODS 449
 The Colorimeter.

CHAPTER LXXIII. THE DESIGN OF A COLOUR WORKS 457

CHAPTER LXXIV. COMMERCIAL NAMES OF PIGMENTS 460

APPENDIX.

THE CONVERSION OF METRIC INTO ENGLISH WEIGHTS AND MEASURES—
 CENTIGRADE AND FAHRENHEIT THERMOMETER SCALES . . . 469

INDEX.

CHAPTER I.

INTRODUCTION.

It is doubtful whether another branch of applied chemistry is recorded of so great an age as the colour industry; at the present time there is hardly a race on the face of the earth which does not make use of colours in some form, either for the decoration of their persons or surroundings. The art of preparing colours is as ancient as their use. It is true that we find from the most remote historical records that the so-called earth colours were almost solely employed, and principally those which exist ready formed in nature. But these natural colours also require their particular process of preparation before they fulfil their object, even though this be merely a mechanical operation, such as powdering or levigating. That the oldest nations of whom we possess lasting records, either written or otherwise, really understood the preparation of colours by chemical processes is shown by the common occurrence in the Egyptian mural pictures of figures clad in brightly coloured garments, a proof that the Egyptians not only understood the science of colour manufacturing, but also the more advanced art of fastening colours upon fabrics—dyeing.

The writings of the ancient Greeks, and in part also the scanty remains of their buildings, prove to us completely that they understood the use of colours to such an advanced degree that they already employed them for pictures as works of art. That the Greeks were also acquainted with the preparation of colours and dyeing follows from various passages from the classical writers, in which magnificently decorated rooms and beautifully coloured garments are often described.

Among the Romans, who were the pupils of the Greeks in the arts and manufactures, the prodigal luxury which existed in Rome, especially under the emperors, caused a great demand for colours, which were used in the most profuse manner for the decoration of house and attire. The Roman colour makers had advanced so far in their art that they could colour the human hair rose red.

A glance at East Indian fabrics and pictures, or at the ancient Chinese buildings, whose colouring is a matter of marvel to-day, shows that the Oriental were not behind the Western nations in the discovery of colours and the art of manufacturing them.

In so old an industry it is not remarkable that great changes have taken place in the course of time. The thousands and thousands of experiments made by the alchemists in the attempt to prepare gold failed in their main object, but the tremendous expenditure of time and trouble in this work was not fruitless; upon the great mass of chemical facts discovered by the alchemists were laid the foundations of scientific chemistry. We find on reading the writings of the alchemists that the colour industry is indebted to them for an immense number of its products; the reason being that the alchemists worked by preference on metals, earths and mineral compounds, and from these substances a large number of colours are obtainable, of which many are still in use to-day, and, on account of their cheapness, will continue in use.

The period in which the painters were also the col

makers lies not far behind us. The preparation of many a colour of particular beauty was treated by the fortunate owner of the recipe as a great secret. It was sold by him at a great price. What a difference between that time and the present! There is now no painter among civilised races to whom it would occur to prepare his own colours; the chemical works provide them for him at a low price and in such a condition that they can be immediately used for painting. The Italian painters prepared the highly prized blue pigment, ultramarine, by laborious toil from the costly *lapis lazuli;* to-day, this same colour, more beautiful and deeper in hue, is made by several works, and sold at a price which bears no comparison with that of the colour obtained from the mineral. The latter was worth many times its weight in gold: a pound of the finest ultramarine can now be bought for a shilling or two.

We find a similar comparison in the case of the fine scarlet pigment known as vermilion: formerly the natural vermilion, cinnabar, was sold at a very high price; at the present time the finest vermilion, prepared artificially, can be bought at a low rate. It is no longer necessary for any one to use natural Chinese vermilion as an artists' colour.

Whilst formerly mineral colours were used in great preponderance, we now know a great number of vegetable and animal colouring matters. The discovery of the sea route to India and the discovery of America had an important influence in this development. From these countries, as from other tropical lands, come the majority of the plants which contain colouring matters. The attempt to change these colouring matters into insoluble compounds led to the discovery of the lake pigments.

With the advance of chemical knowledge the number of colours grew apace; *e.g.*, the discovery of chromium was of great importance to the colour industry: it presented us with a large number of new colours. To a more limited extent, the discovery and study of uranium, molybdenum and other metals were the occasions for the invention of new colours.

In more recent times, efforts in the colour industry have been especially directed to making colours more permanent and, at the same time, harmless. In the first respect, the position at present leaves much to be desired; but, as regards the second property, great advances have been made. The colours in use in former days were almost all very poisonous compounds; the greater number were derived from lead, copper, mercury or arsenic. More recently these poisonous substances have been in many cases replaced by innocuous materials, so that among the colours now in use, though the list is much more comprehensive than of old, there are but few poisonous to a high degree.

In all civilised states the use of poisonous colours has been much restricted by law, and in those cases in which an article is to be manufactured for use as food the employment of such colouring matters has been absolutely forbidden. For example, in Germany by the law of 5th July, 1887, concerning the use of dangerous colours in the preparation of foods and condiments, the application of the permissible colours has been exactly defined.

During the last decades the colour industry and, still more, dyeing have undergone a complete change. The momentous discoveries which have been made in these departments leave far behind the advances which have been made in other branches of chemical technology, the manufacture of explosives, perhaps, excepted. We allude here to the beautiful colours which have been made from coal tar, colours which far surpass in beauty all hitherto known, and which we can already prepare in every shade and hue. Unfortunately, we can only employ the coal-tar colours, as such, in a restricted measure among the pigments; they are of more importance in dyeing. We use the term pigments here in the narrow sense of such substances which, when spread out on certain materials, provoke a certain sensation of colour. Dyeing is, on the contrary, that branch of colour chemistry which generally has for its object the simultaneous production of the colour and its fixation upon a fabric. This definition was at least applicable to the majority of the colours which were in use before the discovery of the coal-tar colours and their introduction into the industry. Since, however, the latter have acquired so great a preponderance in dyeing, it is no longer applicable, for the dyers use at present a large number of substances which are included in the narrow definition of pigments. The greater part of the coal-tar colours are substances which, in solution, when brought in contact with a fabric, adhere to it and colour it permanently.

According to their use and preparation, pigments are divided into a number of classes, and one speaks of painters', artists', enamel, porcelain and glass colours, also of oil, honey, water and cake colours. Although this division is important for trade purposes, it is of little moment for the colour maker, for he can prepare the same colour for both purposes, either for oil or water-colour. What is of the greatest interest for the colour maker is the preparation of the pigment itself. The conversion of the prepared pigment into (oil or water) paint is unaccompanied by difficulties.

When we look for a practical classification for pigments, we find that there are colours which exist ready formed in nature, and others which can only be obtained by certain chemical processes, at times very complicated.

As regards the first group of pigments—those which exist ready formed in nature—the processes which they undergo at the hands of the colour maker are almost entirely mechanical treatments—grinding, sieving, levigating and similar operations—in order to convert them into such a condition that they can be used for painting. Since a large number of these pigments belong to that class of minerals which mineralogists call earths, these pigments have also been designated earth pigments, a term which we shall retain on account of its general use, although it is incorrect, since many of the so-called earth pigments are not obtained from "earths" in the mineralogical sense.

Among the pigments which are prepared by human skill many divisions can be drawn. A large number of pigments are prepared from mineral sources; an equally important number are derived from the animal and vegetable kingdom, the latter consisting of combinations of organic materials with certain inorganic substances. Some few pigments (putting aside the coal-tar colours) are simply organic products, as, for example, the majority of the black pigments, which consist of carbon.

The following classification is drawn up on the lines indicated above:—

1. *Natural Colours or Earth Pigments.*—Found ready formed in nature and requiring only mechanical preparation to be usable. A large number of handsome and also cheap colours belong to this class.

2. *Artificially Prepared Mineral Pigments.*—Obtained by certain chemical processes, and, according to their composition, either compounds of metals with sulphur, oxygen, iodine, cyanogen, etc., or of oxides with acids, *i.e.*, salts.

3. *Lakes.*—Compounds of colouring matters from the animal or vegetable kingdom with a mineral substance, such as lead oxide or alumina.

As a fourth group we might take those colours which do not fall into the previous classes, as, for example, the black pigments composed of carbon; but since this division is not made in practice we shall not regard this species of pigment as a particular group, but shall discuss them in the proper place.

As an entirely new group of colours are to be classed those which are generally called coal-tar colours. These colours, which, at present, are the most important in dyeing and calico printing, are prepared from so-called organic compounds (more properly, carbon compounds). The manufacture of these colours is a separate branch of chemical industry.

CHAPTER II.

THE PHYSICO-CHEMICAL BEHAVIOUR OF PIGMENTS.

IN a work which, as its title indicates, is devoted to a description of the manufacture of pigments, the properties of those substances which are necessary for the preparation of colours cannot be exhaustively considered; we must, therefore, presuppose a knowledge of the elements of chemistry. We have to consider in this book the chemistry of colours; the reader will, therefore, not expect an exposition of general chemical laws; we shall only state certain facts which are of value to the manufacturer. With the description of the manufacture of each pigment and of the materials required for that manufacture, we shall still discuss the chemical processes which must be conducted in the preparation of the colours, so far as it is necessary in order to understand them. In this chapter we shall say a few words about the physical and chemical behaviour of pigments in general.

The great majority of pigments are prepared by the process of precipitation, generally by mixing the solutions of two substances, upon which an interchange of the constituents occurs and the less soluble compound separates in pulverulent form from the solution as a precipitate. Most of these colours are obtained by the admixture of the solutions of two salts; the preparation of the so-called chrome yellow may be taken as an example. In the preparation of this pigment, a solution of a lead salt, sugar of lead (lead acetate), is mixed with a solution of bichromate of potash, whereupon a precipitate of lead chromate (chrome yellow) is formed, whilst potassium acetate remains dissolved. The lead chromate is formed because the acetic acid has a greater affinity for potash than for lead oxide, wherefore an interchange of acid and base takes place, but the lead chromate being insoluble in water consequently separates in the form of a precipitate.

Many mineral pigments are produced in the form of precipitates by passing sulphuretted hydrogen or carbonic acid gas into certain metal solutions. In these cases a similar exchange takes place between the reacting substances to that given in the case of chrome yellow; the metals have a greater affinity for the sulphur or for the carbonic acid than for the substances with which they are already united, they unite with the former, and the new compound separates as an insoluble substance. We have examples of such compounds in cadmium sulphide, which is obtained by passing sulphuretted hydrogen into the solution of cadmium in an acid, and in white lead, which is formed by the saturation of a solution of lead acetate by carbonic acid.

Many organic colouring matters, soluble in water, have the property of forming compounds with metallic oxides, soluble with great difficulty, when their solutions are mixed with a salt of lead, tin or aluminium, and the oxide is separated from the solution by an alkali. The precipitates obtained in this way are insoluble compounds of the colouring matter and the oxide of the metal; they are called lake pigments, or, briefly, lakes. A large number of pigments, often of great beauty, is obtained in this manner. The lakes are widely used in all branches of painting and dyeing.

Of great importance for the quality of the pigment is the physical condition of the precipitate; this is either crystalline or amorphous, that is, non-crystalline. When a crystalline precipitate is examined under the microscope, it is seen to consist of very small, coloured, transparent crystals. The amorphous precipitates are, however, in such a fine state of division that even with the highest magnification they transmit little or no light, and consequently appear opaque. These different characters of precipitates have the greatest influence on that property of pigments which we call *covering power*. In consequence of its transparency, a crystalline precipitate will allow the colour of the surface upon which it is spread to appear through, hence it must be laid on much more thickly than is necessary with an opaque pigment, of which a thin coating is sufficient to make the colour of the surface beneath invisible.

How extremely important is the crystalline or non-crystalline nature of a precipitate in practice is seen by a consideration of white lead. This pigment, lead carbonate, can be made by mixing solutions of a lead salt and a soluble carbonate (soda); but in this case a lead carbonate of crystalline nature is formed, which, being transparent, is of so small covering power that this process has no application in the manufacture of white lead; but a far more troublesome method is used by which a non-crystalline product, amorphous lead carbonate, is obtained.

Many pigments are formed by burning (oxidising) metals, as, for example, zinc white; others are prepared by melting salts together, as Naples yellow; others again are formed by very complicated processes still partially unexplained, as is the case with ultramarine. In the manufacture of colours we find all chemical processes in use.

It may be here remarked that it is quite possible to manufacture some colours, indeed a large number, according to fixed directions, without any particular chemical knowledge being necessary to carry on the processes. Indeed, in works we find most processes being carried out by ordinary labourers who are quite destitute of any knowledge of chemistry. We must, however, add that we are convinced that any colour maker who works simply in a purely empirical manner, according to a stereotyped recipe, will never be in the position to raise himself above the position of a workman; he will not be able, when a slight mishap occasions a change in the ordinary course of the process, to devise a means of overcoming the defect, but will be compelled to dispose of the faulty product in the condition in which it exists. Such a manufacturer is in a condition of blind dependence on the chemical works and dealers from whom he receives the raw materials requisite for the preparation of his colours. If he should receive materials which contain impurities not to be detected by empirical methods, the inevitable result will be that the colours produced from them will not be equal to the standard. If, in making a colour which is the outcome of several processes, a workman once makes a mistake, the product will not be of the required quality.

On the contrary, if the manufacturer possesses a certain amount of chemical knowledge, it will not be difficult for him to ascertain the causes of a failure in a process, and, at the same time, to devise means by which the defects may be removed. The manufacturer is more and more in the habit of buying the chemicals which he requires for his manufactures rather than of making them himself. He should, therefore, be in a position to form an opinion as to the usability and purity of these substances, which will only be possible when he has the knowledge requisite for subjecting them to a chemical examination.

Although we shall presuppose, as we have said, that those who intend to concern themselves with colour manufacturing possess an acquaintance with the principles of chemistry, yet this book has been so planned that it may be of use (we hope) to the practical man who is innocent of chemical knowledge. On this account, we have devoted care to the description of those raw materials which are bought in large quantity, and to the simple investigation of their purity.

When the manufacturer has the advantage of a chemical education, apart from his endeavours to produce colours lacking nothing in beauty or depth of shade, he will direct his endeavours in two directions, in respect of which great advances are yet to be made—the permanence and harmlessness of his colours.

Many pigments possess the undesirable property of losing their brightness under atmospheric influences; many, indeed, fade away completely in the course of time. We have only to examine a picture some centuries old; in spite of the care bestowed on its preservation, we can say with certainty that, in the course of time, it will be so completely altered that nothing will remain of the original colours. It is the endeavour of the sensible manufacturer of colours to make only such as remain unaltered by atmospheric action, and also undergo no change when they are mixed with other pigments. Although it may be highly desirable that the painter should possess a knowledge of the chemical properties of the colours he uses, still it should be the first object of the maker to take care that he places on the market only colours which will remain as much as possible unaltered when used alone, and will remain undecomposed when mixed. This is, unfortunately, not the case with many colours now in use. We shall return later to this point, of such extraordinary importance to the artist.

The second point to be observed, is to produce only harmless colours. The advances of chemistry have made known to us a series of colours which have the advantage over others known for a longer period that they are non-poisonous.

Unluckily, these harmless colours frequently fall behind the poisonous colours in brilliance, and generally they are more expensive. Here, too, is opened to the manufacturer a wide field of activity. The more completely poisonous substances disappear from the colours in use, the more widespread will be the use of colours. We should remark that the expression "poisonous colours" is to be used with a certain reserve. Many pigments which contain lead, copper, antimony, mercury, etc., are poisonous, because they contain poisonous metals; but poisoning with them will not readily take place on account of their insolubility. It is different with the very poisonous arsenic compounds, which should be removed from the list of colours in common use; many a misfortune caused by them would then be avoided.

Endeavours to produce innocuous colours have been more successful than the efforts after permanence. There are now very few commonly used colours which can be accounted very poisonous compounds, and which cannot be replaced by other colours of equal beauty. On the whole, we are now in the position to prepare harmless colours suited to most purposes. Special endeavours should be made to sell these, so that such cases of poisoning should not occur as, for example, caused by gingerbread which had been wrapped in paper coloured by emerald green.

CHAPTER III.

RAW MATERIALS EMPLOYED IN THE MANUFACTURE OF PIGMENTS.

As we have mentioned before, the manufacturer of colours now generally uses materials supplied to him by chemical works. The purer these are, the easier it will be to work with them, and the finer will the colours turn out. We have indicated that it is important for the manufacturer to know accurately the properties of his materials in order to be able to estimate their value. Many substances required in certain cases must be made by the colour manufacturer, since, on account of their condition, they cannot form articles of commerce—chlorine and sulphuretted hydrogen, for example.

In addition to the substances which are not to be bought, there are others which do occur in commerce, but are sold at so high a price that the manufacturer is compelled to make them himself. This is the case with the cobalt compounds, from which many beautiful colours are made. The producers of these demand such prices that it is to the interest of the colour maker to prepare them for his own use.

In the following chapters, we shall deal with the more important raw materials which are employed in colour manufacturing, and shall restrict our remarks to what is of particular importance thereto. For more detailed accounts of these raw materials the reader is referred to the text books of chemistry, in which he will find them minutely described, in so far as they are chemical products.

The materials employed may be divided into assistants in the processes and components of the manufactured pigment. The assisting substances are those which are used in the manufacture of a colour without entering into its composition; from the component materials the colours are directly derived. For example, in the manufacture of Prussian blue, yellow prussiate of potash, an iron salt, water (in which the salts are dissolved) and nitric acid are used. In the blue obtained are contained portions of the iron salt and of the yellow prussiate, these are, therefore, component materials, whilst water and nitric acid are simply assistants, since they do not enter into the composition of the pigment.

In colour making a large number of assisting materials are employed, which comprise a considerable number of elements and compounds. Since these are of great importance for our purpose, we shall describe their properties, and, when necessary, briefly the method of preparation.

Among the component materials are to be reckoned a large number of salts of the alkaline earth and earth metals and of all the heavy metals. In addition, there are also the substances of animal or vegetable origin used in lake making.

In the description of the raw materials, if we were to overstep the line drawn here, we could include a great variety of compounds, those, for example, used in the manufacture of the so-called aniline dyes. These substances form, however, as we have stated, the object of a particular branch of manufacture, which forms a separate division of colour chemistry, but with which is not to be confounded what has been hitherto designated the manufacture of colours.

CHAPTER IV.

ASSISTANT MATERIALS.

Water, $H_2O = 18$.[1]—This substance plays a tremendous part in colour making; almost all the substances which are used in solution are dissolved in water; the removal from precipitates of admixed foreign bodies, the so-called washing, is always accomplished with water. The chemist does not understand by water quite that liquid which in general speech is so designated. We must consider the water which is at the disposal of the colour maker.

Water, in the chemical meaning of the word, is a liquid composed only of hydrogen and oxygen, and leaving no residue when evaporated. Such water is not found in nature; it can only be obtained by distillation of well or river water. The water which falls in long continued rain, or is obtained by melting snow, is most nearly like distilled water; it contains only small quantities of dissolved substances, and generally such as would be without influence in colour making. Water of this description is available for but a limited use; the large quantities of water required in a colour works must be taken from springs or streams. These waters contain, however, more or less large amounts of dissolved salts, which act in a marked manner upon the substances dissolved in them.

In almost all spring and well waters is found carbonate of

[1] We append the chemical formula and the molecular weight to the description of each compound.

lime; such waters are called "hard". River water contains generally little carbonate of lime; it is then called "soft water". The influence of the carbonate of lime is especially evident when salts of lead, copper, iron and other heavy metals are dissolved in water; the carbonate of the particular metal gradually separates from solution, and the liquid becomes very turbid.

When only hard waters containing much lime are at the service of the manufacturer, turbid solutions are often obtained, which must be filtered before use. In many cases this can be avoided by adding milk of lime to the water in a large vessel; the free carbonic acid unites with the lime, and thus the carbonate of lime, which is only soluble in water containing free carbonic acid, separates as a fine precipitate. Water which has been treated in this way becomes clear after some time, through the deposition of the carbonate of lime; it is then soft water. In order to separate the carbonate of lime in this way, no more than the requisite quantity of milk of lime should be added, so that no lime remains in excess, since this would cause precipitates when salts of lead, copper, iron, etc., were dissolved. In many cases—for example, when lead or barium salts are dissolved—the lime contained in the water can be made harmless by slightly acidifying with acetic or nitric acid. Water which contains sulphate of lime (gypsum) is equally useless for many purposes, as, for example, the solution of lead and barium salts. These metals form insoluble compounds with the sulphuric acid, which render the solution turbid, and can be removed only with difficulty by filtering, on account of their great fineness. They are more easily removed by allowing to settle.

Water containing gypsum often contains in addition small quantities of sulphuretted hydrogen. However small the quantities of this gas may be, they still make the water absolutely useless for certain purposes in the manufacture of colours; for example, for the preparation of all pigments containing lead which are obtained by precipitation. The sulphuretted hydrogen forms black compounds with lead, copper, bismuth, mercury and other metals, which impair the brilliance of the colour. A colour made under these conditions is never clean, its hue is injured by the admixture of the black substance.

Water which contains much common salt (sodium chloride) is unsuitable for the solution of lead, mercury and silver salts. In consequence of the great affinity of these metals for chlorine, turbid solutions are obtained when their salts are dissolved in water containing common salt.

Some waters contain a considerable quantity of iron. Such waters deposit on evaporation, and often on standing exposed to the air, a brown powder of ferric hydrate, which would have considerable influence on the shade of a pigment. White pigments, in the preparation of which such a water is used, have always a brownish tinge; yellow and red pigments are also unfavourably affected.

Carbonate of lime and common salt occur in small quantities in every well water. The colour maker must do the best he can with such a water; its use will not particularly harm the shade of the colours prepared with it if the amount of the impurities is not very large. Water containing much iron is practically useless; the oxide of iron would injure the colours so much that it would not be possible to obtain brilliant shades. Water from wells in the neighbourhood of deposits of turf or cemeteries often contains considerable quantities of organic substances which act injuriously on the shade of pigments; such water should not be used in colour making.

The impurities in a water are more or less harmful according to the purpose for which it is to be used. Sulphate of lime is generally more injurious than carbonate of lime, since the precipitates which the latter causes in solutions of the salts of certain metals can be prevented by the addition of acids. This is not the case with sulphate of lime; when lead or barium salts are dissolved in water containing this substance, a precipitate of lead or barium sulphate is obtained, which is insoluble.

In dealing with the salts of costly metals, such as mercury or silver, it is better to dissolve them in distilled water, or, at least, very pure rain water. The rain water which runs from zinc or well tiled roofs is generally very pure; for practical purposes it may be regarded as free from carbonate and sulphate of lime, sulphuretted hydrogen and common salt. The colour maker should take care to obtain as much of this pure water as possible by erecting large rain-water tanks.

The less impurity a water contains the more useful it is for our purpose. After rain water soft river water is the best, and after this the softer well waters. All mineral waters distinguished by a high content of salts or gases are quite useless for colour making; for this reason sea water is disqualified.

An accurate analysis of a water is much too complicated for the manufacturer; it is sufficient for him to convince himself of the absence of certain substances. Water which, some time after the addition of a little tannic acid solution, acquires a clear green or a bluish to black shade contains much iron, and is useless. Water which coagulates a large quantity of a solution of soap in alcohol is very rich in carbonate or sulphate of lime. In order to decide approximately in what relative proportion these salts are present, a solution of barium chloride is added to the water so long as a precipitate forms. If this disappears completely on the addition of nitric acid, the water contains only carbonate of lime; if it only partly dissolves, sulphate of lime is also present. The presence of chlorine is shown by a considerable turbidity on acidifying the water with nitric acid, boiling and adding silver nitrate. If the precipitate obtained on the addition of a lead salt is not pure white, but discoloured, the water contains sulphuretted hydrogen, which has formed black lead sulphide. In order to test the water for organic substances, about a litre is evaporated to dryness in a porcelain dish, and the residue heated to redness; if it turns brown and black, and possibly gives off a smell of burnt feathers, the water contains much organic matter.

Pure water is coloured permanently red by a solution of potassium permanganate; but if it contains organic matter, the solution is decolourised after some time and a brown precipitate is deposited at the bottom. From the amount of this precipitate an idea of the quantity of organic matter present may be obtained.

It is only necessary to be very scrupulous concerning the quality of the water when it is to be used for the solution of salts or the extraction of dye-woods. For washing precipitates, which requires a large volume of water, there can generally be used, without detriment, water containing much lime, but it must be free from iron and sulphuretted hydrogen. The latter is particularly harmful to most of the lead colours, which would lose in beauty by washing with water containing this substance.

It is hardly necessary to say that the water used in colour making must be quite clear. Muddy river water must in every case be completely freed from the solid particles contained in it, either by settling or by filtering. Filters filled with well-washed sand give good results for this purpose.

Chlorine, $Cl = 35\cdot5$.—For some operations in colour making it is necessary to employ chlorine. This is a greenish yellow gas at ordinary temperatures, which is characterised by a suffocating smell and the energy with which it unites with most elements. On account of its injurious effects on man certain precautions have to be observed in preparing chlorine, and it is advisable to erect the apparatus necessary for its production in a separate room, so that the workmen are not injured by the gas.

Formerly chlorine was exclusively made in lead apparatus, because this metal is one of the least readily attacked. When such an apparatus is used for the first time a layer of lead chloride is formed, which, like a varnish, protects the metal beneath from further attack. Fig. 1 represents an apparatus

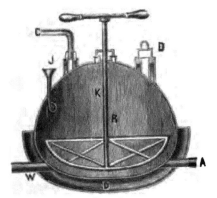

FIG. 1.

formerly employed for the preparation of chlorine in chemical works. In the upper part of the pear-shaped vessel, K, there are four openings, two of which, D and C, are provided with water lutes. This means that the opening is surrounded by a moat containing water, into which the rim of the cover dips, thus making a joint. Through the middle opening goes the axle of the stirring apparatus, R; in the fourth is a lead safety funnel, J. Solid materials are introduced through D, liquid through J; the tube C carries away the chlorine formed; the tube A, furnished with a stop cock, can draw off the fluid contents of the apparatus.

Since lead melts at low temperature, the apparatus cannot be heated over the fire without danger, therefore it is surrounded by an iron jacket, W, which is filled with water, or else the apparatus is heated by steam introduced into W. Larger quantities of chlorine are more conveniently prepared in an apparatus, of similar structure, made of stone or

FIG. 2.

earthenware, which have the advantage over lead that they are not at all attacked by chlorine.

Fig. 2 exhibits the construction of such an apparatus of medium size. It is constructed of sandstone or earthenware; the lid and some of the smaller parts can be made of either earthenware or lead. The pyrolusite is introduced at G in large pieces; H is the funnel for pouring in the acid;

$E\ K\ D$, the steam pipe; C, the perforated false bottom upon which the pyrolusite lies; F, the delivery tube for the chlorine; J, the opening for running off the manganese chloride; a, the leaden cover.

To prepare chlorine, 1 part (by weight) of common salt, 1 part of powdered pyrolusite, $2\frac{1}{2}$ parts of vitriol and $1\frac{1}{4}$ part of water are used. The salt and the pyrolusite are introduced through D into the apparatus (Fig. 1); the acid, diluted with the water, is poured in through the funnel, the materials are mixed by the stirrer and gently warmed until chlorine appears, when the application of heat must be considerably diminished or the chlorine will be violently evolved.

Pyrolusite and hydrochloric acid are now generally used for the preparation of chlorine, because the solution of manganese chloride, left at the end of the operation, is valuable.

If all the chlorine made in one operation is not at once required for the manufacture of a colour, it can be utilised by sending it into a box filled with slaked lime, which is converted into chloride of lime or bleaching powder. The liquid run away from the apparatus at the conclusion of the operation contains manganese and sodium sulphates, or manganese chloride as the case may be, and can be used for the preparation of manganese pigments.

Ammonia, $NH_3 = 17$. — Ammonia is obtained from chemical works in the form of a strong solution of ammonia gas in water, which is generally very pure. The density of an aqueous solution of ammonia is the smaller the more ammonia it contains, and thus the strength of a solution of ammonia can easily be formed by means of the hydrometer. The following table shows the percentage of ammonia, NH_3, in a liquid of known specific gravity at the temperature of 14° C. :—

Specific Gravity.	Ammonia per cent.	Specific Gravity.	Ammonia per cent.
0·885	36·00	0·953	11·50
0·886	35·00	0·955	11·00
0·889	34·00	0·957	10·50
0·891	33·00	0·959	10·00
0·893	32·50	0·961	9·50
0·895	32·00	0·963	9·00
0·897	31·30	0·965	8·50
0·900	26·00	0·968	8·00
0·905	25·39	0·970	7·50
0·925	19·54	0·972	7·00
0·932	17·52	0·974	6·50
0·947	13·46	0·976	6·00
0·951	12·00	0·978	5·50

The Hydrometer.—In the above table the percentage content of the ammonia solution is given according to its specific gravity, that is, according to the ratio between the weight of any volume of the liquid and the weight of an equal volume of water. According to scientific principles, only those hydrometers should be used which are graduated in specific gravities. In spite of all exertions in this direction, manufacturers have not yet been induced to use such instruments in every case. Hydrometers, with quite arbitrary scales, such as those of Baumé and Twaddell, are frequently found in works. These hydrometers generally only show that a liquid is of so many degrees on the particular scale, and the manufacturer in using them is restricted to the following out of a certain recipe which requires the use of a liquid of a certain strength which is expressed in degrees Baumé, etc. He does not learn by this how many per cent. of the particular substance are dissolved in the water when the liquid has a certain hydrometric strength.

For the sake of uniformity, it is urgently to be desired that all manufacturers who use the hydrometer to estimate the content of a liquid in ammonia, potash, soda, hydrochloric, sulphuric, nitric acids, etc., should employ simple specific gravities. This is desirable, because the percentage strength of a solution, corresponding to the specific gravity, can be at once accurately found from tables. On these grounds, in the present work, we have restricted ourselves to tables showing simply the specific gravities of solutions and the corresponding composition.

Sal Ammoniac or Ammonium Chloride, $NH_4Cl = 53·5$.— This substance comes into commerce in the form of a white crystalline meal, more rarely in the form of sugar loaves (crystallised sal ammoniac) or of flat cakes (sublimed sal ammoniac). It is usually very pure, since impure forms, generally containing much iron, are difficult of sale. At a particular temperature sal ammoniac is volatile; it is used in certain mixtures in order to prevent the temperature, on heating, from rising beyond a certain point. Like ammonia, it is more used in dyeing.

Ammonium Sulphide, NH_4HS.—This compound is obtained by leading sulphuretted hydrogen into ammonia solution so long as it is dissolved, and a test portion of the liquid still gives a white precipitate with a solution of magnesium sulphate. Ammonium sulphide decomposes by long standing in the air, sulphur being separated. It gives precipitates with the salts of certain metals, for example, iron, cobalt, manganese, zinc, nickel. These precipitates, which consist of the sulphides of the metals, are not formed by sulphuretted hydrogen in acid solutions.

Acids.

In colour making many acids are used for the solution of metals, the production of precipitates, for oxidations and so forth. Commercial acids, especially inorganic acids, generally contain not inconsiderable quantities of impurities which are injurious in the manufacture of many colours.

Hydrochloric Acid, $HCl = 36·5$.—The commercial acid (muriatic acid, spirits of salt) generally contains large quantities of iron, which colour it yellow—fortunately, in many cases, this is not a disadvantage, and also at times the iron can be removed from solutions made in the acid. Another impurity is sulphuric acid. This can be detected by diluting and adding barium chloride; if sulphuric acid be present, a white precipitate, or, at least, a cloudiness, appears.

Ordinary hydrochloric acid is a solution of hydrochloric acid gas in water. The strongest acid contains 42·85 per cent. of the gas, and has the specific gravity 1·21. The following table gives the strengths of acids of various specific gravities:—

Specific Gravity.	Hydrochloric Acid per cent.	Specific Gravity.	Hydrochloric Acid per cent.
1·21	42·85	1·10	20·20
1·20	40·80	1·09	18·75
1·19	38·88	1·08	16·71
1·18	36·36	1·07	15·49
1·17	34·34	1·06	13·86
1·16	32·32	1·05	11·49
1·15	30·30	1·04	8·97
1·14	28·28	1·03	6·93
1·13	26·26	1·02	4·89
1·12	24·24	1·01	2·03
1·11	22·22	—	—

Sulphuretted Hydrogen, $H_2S = 34$. — This is a gas of acid properties smelling like rotten eggs; it precipitates the sulphur compounds of many metals when led into the acid solution of the corresponding salt. This substance is seldom required in colour works, so that it is convenient to have an apparatus which permits of the preparation of any required quantity. Fig. 3 represents an apparatus devised by the author, which is well adapted for the preparation of sulphuretted hydrogen. It consists of a small, wooden tub, on whose upper edge lies a thick paper ring, so that the lid may be pressed down air-tight by the screws B. Through the lid pass a tap-funnel, T, a movable screw, S, and a tube, R, to carry away the gas. On the screw S hangs a basket, K, by a handle; this is filled with pieces of iron sulphide as large as nuts. The tub is filled to about one third of its height with a mixture of 9 parts of water and 1 part of sulphuric acid.

When sulphuretted hydrogen is required the basket is lowered by the screw, S, until it dips in the liquid; according as the basket dips more or less into the liquid a fast or slow current of the gas is obtained. When the gas is no longer

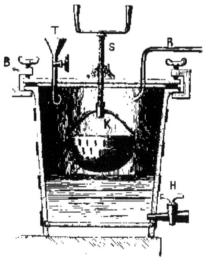

Fig. 3.

required the basket is raised out of the liquid, and the evolution of gas at once ceases. The funnel, T, serves for the introduction of the liquid, the tap, H, for drawing off the iron sulphate solution, which can be used with advantage for the preparation of fine iron colours. The apparatus should not be opened so long as sulphide of iron remains in the basket.

Sulphuric Acid comes into the market in two different forms: oil of vitriol and fuming or Nordhausen sulphuric acid; both are used in colour making. *Oil of Vitriol*, H_2SO_4 = 98, is a colourless, oily liquid of high specific gravity; it is generally tolerably pure, and contains, as a rule, only a small quantity of lead, the presence of which is indicated by a turbidity on largely diluting the acid. The amount of pure sulphuric acid in the liquid is practically determined by taking the specific gravity. The table indicates the relation between the specific gravity and the content of sulphuric acid.

Specific Gravity.	Sulphuric Acid per cent.	Specific Gravity.	Sulphuric Acid per cent.
1·8485	100	1·8043	89
1·8475	99	1·7962	88
1·8460	98	1·7870	87
1·8439	97	1·7774	86
1·8410	96	1·7673	85
1·8376	95	1·7570	84
1·8336	94	1·7465	83
1·8290	93	1·7360	82
1·8233	92	1·7245	81
1·8179	91	1·7120	80
1·8115	90	—	—

Nordhausen Sulphuric Acid, $H_2S_2O_7 = H_2SO_4 + SO_3$, is generally a yellowish brown liquid, which gives off white fumes in the air. It contains varying quantities of sulphur trioxide dissolved in sulphuric acid. It often contains selenium, which separates as a red powder when the acid is diluted. The presence of this impurity does not interfere with the use of the acid for dissolving indigo, the only purpose for which it is required in the colour factory.

Nitric Acid, $HNO_3 = 63$.—This acid, which is used in the preparation of many colours, is distinguished by the readiness with which it gives up part of its oxygen, and thus converts metals like antimony and bismuth into oxides, and transforms other compounds into a higher state of oxidation. There are two kinds of nitric acid: ordinary nitric acid, a colourless liquid which is more or less pure; and fuming nitric acid, a yellow or orange coloured liquid, fuming strongly in the air, which consists of a solution of nitrogen peroxide, NO_2, and nitric oxide, NO, in nitric acid.

Specific Gravity at 15° C.	Nitric Acid per cent.	Specific Gravity at 15° C.	Nitric Acid per cent.
1·530	100·00	1·323	55·00
1·520	97·00	1·284	50·49
1·509	94·00	1·251	45·00
1·503	92·00	1·211	40·00
1·499	91·00	1·185	33·86
1·495	90·00	1·157	30·00
1·478	85·00	1·138	25·71
1·460	80·00	1·120	23·00
1·442	75·00	1·089	20·00
1·423	69·96	1·067	15·00
1·400	65·07	1·022	11·41
1·346	60·00	1·010	4·00

Since the action of nitric acid chiefly depends on its oxidising properties, which are possessed by both kinds, it generally does not matter which is used. The usual impurities are chlorine and sulphuric acid; the presence of the first is shown by silver nitrate solution, of the latter by barium chloride, in each case added after diluting. When the acid is used for oxidations these impurities do not interfere, but nitric acid containing chlorine cannot be used to dissolve silver, because the chlorine would form insoluble silver chloride.

The strength of nitric acid is gauged by its specific gravity as given in the table.

Aqua Regia.—A mixture of 2 parts of hydrochloric acid and 1 part of nitric acid gradually turns orange or yellow and evolves chlorine. This liquid, which can dissolve gold in consequence of the free chlorine it contains (hence its alchemistic name, from gold, the "king of metals"), is used as a very powerful oxidising agent in colour making.

Carbon, $C = 12$, is the only one of the non-metallic elements to be mentioned here; by itself it forms a group of very important pigments, which we shall describe in detail at a later stage.

Carbonic Acid Gas, $CO_2 = 44$, is used in the manufacture of white lead, which it precipitates from lead acetate. This is, however, a particular branch of colour making carried on in special works. In describing this manufacture we shall return to the preparation of carbonic acid on a large scale.

Organic Acids.

The organic acids which are important in colour making are acetic, oxalic and tartaric acids.

Acetic Acid, $C_2H_4O_2 = 58$.—The very dilute form of this substance is known commonly as vinegar, the stronger as pyroligneous acid, and the purest as glacial acetic acid; the latter is, however, scarcely used. Formerly in colour making ordinary vinegar was used, but now pyroligneous acid is almost exclusively employed. This is distinguished by its strong empyreumatic smell, which, however, is without importance in colour making.

The strength of a solution of acetic acid cannot be found by a simple estimation of specific gravity, since the density does not increase with the percentage of acetic acid. If an accurate estimation of the strength of acetic acid is required, it must be obtained by neutralising the acid with an alkali by a process of volumetric analysis.

For practical purposes, where it is generally known whether a very strong or a more dilute acetic acid is under consideration, the following table, showing the connection between specific gravity and percentage strength, is sufficient.

Specific Gravity.	Acetic Acid per cent.	Specific Gravity.	Acetic Acid per cent.	Specific Gravity.	Acetic Acid per cent.	Specific Gravity.	Acetic Acid per cent.
1·0635	100	1·0720	74	1·058	48	1·031	22
1·0670	98	1·0710	72	1·055	46	1·027	20
1·0690	96	1·0700	70	1·054	44	1·025	18
1·0706	94	1·0700	68	1·052	42	1·023	16
1·0716	92	1·0690	66	1·051	40	1·020	14
1·0730	90	1·0680	64	1·049	38	1·017	12
1·0730	88	1·0670	62	1·047	36	1·015	10
1·0730	86	1·0670	60	1·045	34	1·012	8
1·0730	84	1·0660	58	1·042	32	1·008	6
1·0730	82	1·0640	56	1·040	30	1·005	4
1·0735	80	1·0630	54	1·038	28	1·002	2
1·0732	78	1·0620	52	1·035	26	1·001	1
1·0730	76	1·0600	50	1·033	24	—	—

Oxalic Acid, $C_2H_2O_4 . 2H_2O = 126$, has but a limited use in colour making. It comes into commerce in the form of more or less pure white crystals which readily dissolve in water, and are almost pure oxalic acid, containing only small quantities of oxalate of lime, the presence of which is without importance for the purposes to which the acid is put in colour making. Frequently, instead of oxalic acid, the acid potassium oxalate (salt of sorrel) is used.

Tartaric Acid, $C_4H_6O_6 = 150$, occurs as white or yellowish crystals, with a slightly burnt smell, which dissolve readily in water, and have a strong acid taste. The pure acid, which is white and without smell, is considerably dearer than the yellow variety. The impurities of the latter, which are small in quantity, are without influence on the colours prepared by its help, so that this form is generally used.

CHAPTER V.

METALLIC COMPOUNDS.

ALKALIS.

THE compounds of the alkali metals, potassium and sodium, play a considerable part in colour making. Formerly the potassium compounds were in general use, but the sodium compounds are at present obtainable at a much lower price, and in most cases they can be used equally well. Thus, in colour making, sodium compounds are chiefly employed. The cyanogen compounds are an exception; their potassium compounds are used exclusively.

Potassium Compounds.—The potassium compounds which are chiefly used in colour making are potassium carbonate (potashes, pearl-ash), potassium hydroxide (caustic potash), potassium nitrate (saltpetre), potassium tartrate (tartar), and potassium ferrocyanide and ferricyanide (yellow and red prussiate of potash). The cyanogen compounds have peculiar properties. We shall describe them separately after the potassium and sodium compounds.

Potassium Carbonate (carbonate of potash), $K_2CO_3 = 138$, is known commercially as potashes, a name derived from its former method of preparation by heating the ashes of plants in pots. At present potashes are prepared in large quantities from other sources.

Pure potash forms crumbling lumps with a slight yellow or bluish grey tinge, rapidly absorbing moisture from the air, and in time completely liquefying. The yellowish tinge

METALLIC COMPOUNDS.   33

is caused by oxide of iron, the bluish by manganese compounds. The so-called calcined potash has been strongly heated, and thus all organic substances contained in it have been destroyed.

Potashes are in no way pure potassium carbonate; they contain a mixture of all those salts which are found in plants—potassium sulphate and chloride, small quantities of silicic acid, etc. These impurities are rarely harmful, still it is generally necessary to know the percentage of pure potassium carbonate contained in potashes.

Although at present in commerce the strength of potashes is frequently guaranteed, it is still desirable to estimate the strength. It is sufficient for practice to allow a small quantity, say 100 grammes, to stand with an equal quantity of very cold water for some hours, then to filter and pour a similar quantity of water over the residue on the filter. The weight of undissolved substance subtracted from 100 gives with sufficient accuracy the weight of pure potassium carbonate contained in 100 parts of potashes. This method is founded on the fact that potassium carbonate dissolves readily even in cold water, but the other salts with difficulty. This procedure can also be used to obtain pure potassium carbonate from crude potashes; it is only necessary to filter and evaporate to dryness the solution obtained by pouring very cold water on crude potashes.

Potassium Hydroxide (Potassium Hydrate, Caustic Potash), $KOH = 56$.—The commercial variety consists of very deliquescent white lumps, generally containing a large quantity of impurities. On this account caustic potash, or rather a solution of it, is prepared in the colour works.

With this object 11 parts of potash, contained in a tub with an opening at the bottom, are mixed with 100 parts of cold water. Two hours later, the clear solution is run off into a clean iron pan, in which it is heated to boiling. To the boiling solution is added milk of lime prepared from water and 3·5 parts of quicklime. After the liquid has boiled a few minutes, a small portion is filtered and hydrochloric acid added to the clear filtrate; if no effervescence occurs, then all the potassium carbonate is converted into caustic potash. Should effervescence occur, milk of lime is added until a new portion no longer effervesces on the addition of hydrochloric acid. Then the pan is covered with a well-fitting lid, and the cooled liquid, if not required for immediate use, preserved in well-corked glass bottles.

The strength of a caustic potash solution can be found by means of a hydrometer. The following table shows the relation between the specific gravity of a solution and the percentage of caustic potash it contains:—

Specific Gravity.	Caustic Potash per cent.	Specific Gravity.	Caustic Potash per cent.
1·06	4·7	1·28	23·4
1·11	9·5	1·39	32·4
1·15	13·0	1·52	42·9
1·19	16·2	1·60	46·7
1·23	19·5	1·68	51·2

Potassium Nitrate (Saltpetre), $KNO_3 = 101$, consists of large crystals, which quickly dissolve in water. On heating it readily gives up oxygen, and thus finds use as an oxidising agent. In former times, when the colour manufacturer was compelled to make his own materials, saltpetre was of great importance in colour making; at present, when such materials are to be bought at low prices and no colour maker prepares his own, saltpetre is little used.

Potassium Bitartrate, $C_4H_5KO_6 = 188$.—This salt, known as tartar in large crystals, and as cream of tartar in the form of meal, is occasionally used in colour making. It is little soluble in cold water, but more easily in hot. The hot solution is generally used.

Potassium Bichromate (Bichromate of Potash), $K_2Cr_2O_7$ = 295.—This salt is made in special works, by melting chrome iron ore with saltpetre and extracting the mass with water, when a yellow solution of potassium chromate is obtained; to this sulphuric acid is added, which unites with half the potassium, thus leaving potassium bichromate, which is obtained by evaporation of the solution in fine red crystals. These are purified by recrystallisation. At present, in place of the above method, calcium chromate is formed by roasting chrome iron ore with lime; the calcium chromate is then decomposed by a soluble potassium salt, thus forming potassium chromate.

Potassium bichromate is unaltered in air; it dissolves easily in water, and is of great importance in the preparation of many colours, in particular chromium oxide and the lead pigments. The commercial salt generally contains potassium sulphate, with which at times it is intentionally adulterated. The adulteration is detected by dissolving in water, adding half the volume of pure hydrochloric acid, and cautiously and carefully dropping in spirits of wine. A rapid action takes place, which is only assisted by warming when necessary. The red liquid changes to emerald green. If barium chloride be now added, a white precipitate is obtained in the presence of potassium sulphate.

Potassium Sodium Chromate, $KNaCrO_4$ = 279, is also used in colour making. Its solution is made by adding soda to a solution of potassium bichromate so long as an effervescence of carbonic acid occurs, and until the liquid turns red litmus paper blue; the solution of the double salt is yellow.

Chrome Alum, $KCr(SO_4)_2.12H_2O$ = 499.—This salt occurs in commerce as beautiful violet crystals. It is obtained as a by-product in the manufacture of aniline and anthracene dyes, and may often be bought at lower prices than other chromium salts; 100 parts of water dissolve approximately 20 parts of chrome alum.

Potassium Ferrocyanide, $K_4Fe(CN)_6 \cdot 3H_2O = 422$.—The potassium iron cyanogen compounds are made in special works, particularly in the neighbourhood of large towns, by melting potashes with nitrogenous organic substances and iron, washing out the mass and purifying the salt so obtained by recrystallisation. Potassium ferrocyanide (yellow prussiate of potash) forms large transparent crystals of a peculiar soft nature, which dissolve readily in water. It often contains considerable quantities of potassium sulphate, up to 5 per cent., and it is to be noted that the impurity is much the cheaper of the two salts. When barium chloride is added to a solution of the salt, a white precipitate forms if sulphate be present.

The behaviour of yellow prussiate towards iron salts is noteworthy. With ferrous salts, for example green vitriol (copperas), it gives a white precipitate which gradually turns blue in the air; with ferric salts, for example ferric chloride ("nitrate of iron"), it at once gives a blue precipitate.

Potassium Ferricyanide (Red Prussiate of Potash), $K_3Fe(CN)_6 = 329$, is obtained by passing chlorine through a solution of yellow prussiate until the liquid smells strongly of chlorine and no longer gives a precipitate with a solution of a ferric salt. The solution then contains potassium ferricyanide and chloride. The former is obtained by evaporating and allowing to crystallise.

Pure potassium ferricyanide forms beautiful dark red crystals, which readily dissolve in water. The solution gives a blue precipitate with ferrous salts, but only a brown colouration and no precipitate with ferric salts. Both yellow and red prussiate are used in the preparation of several much-used colours, for Prussian and Chinese blues, and several others. All cyanogen compounds, with the exception of yellow prussiate, are extremely poisonous. The following table gives the solubility of potassium ferricyanide at different temperatures:—

100 Parts of Water dissolve Parts of Salt.	Temperature. °C.	Specific Gravity of Solution.
33	4·44	1·151
36	10·00	1·164
40·8	15·50	1·178
58·8	37·80	1·225
77·5	100·00	1·250
82·6	104·40	1·265

Sodium Salts.—In chemical properties the sodium salts are very similar to the potassium salts, and, being cheaper, they are generally used in place of the latter.

Sodium Carbonate (Soda Crystals), $Na_2CO_3.10H_2O = 286$, is made in enormous quantities in great works and in a very pure state. It forms large transparent crystals, which effloresce in the air, losing a large quantity of water, and so falling to a white powder. Although this property does not interfere with the use of soda, since it is generally used in solution, yet efflorescence should be as far as possible avoided by keeping the salt in well-closed packages, because effloresced soda dissolves more slowly than crystallised, since it has to combine with water before it can enter into solution.

In the retail trade a form of soda is found which is adulterated with very large quantities of Glauber's salt. This is recognised by the different form of the crystals. Manufacturers sell soda stating its strength. The colour maker should only buy with this guarantee.

Sodium Hydroxide (Sodium Hydrate, Caustic Soda), $NaOH = 40$, comes into commerce in the form of hard masses, the use of which would be very convenient for the colour maker if it were not often very impure. Thus it is better to prepare a solution oneself, which is accomplished in the manner given above for caustic potash.

Caustic soda and caustic potash have similar properties; they have a corrosive action on the skin, readily unite with carbonic acid from the air, and separate the heavy metals from their solutions in the form of hydrated oxides.

The strength of caustic soda solutions is given in the following table:—

Specific Gravity.	Caustic Soda per cent.	Specific Gravity.	Caustic Soda per cent.	Specific Gravity.	Caustic Soda per cent.	Specific Gravity.	Caustic Soda per cent.
2·00	77·8	1·56	41·2	1·40	29·0	1·23	16·0
1·85	63·6	1·50	36·8	1·36	26·0	1·18	13·0
1·72	53·8	1·47	34·0	1·32	23·0	1·12	9·0
1·63	46·6	1·44	31·0	1·29	19·0	1·06	4·7

Besides soda and caustic soda, few soda salts are used in colour making. However, sodium nitrate (Chili saltpetre), $NaNO_3$, is frequently used instead of ordinary saltpetre, from which it differs in being deliquescent.

Sodium Thiosulphate (Hyposulphite), $Na_2S_2O_3.5H_2O$ = 248, is a common article of commerce, being much used by photographers; it is used in a few cases in colour making. It forms large crystals with a somewhat bitter taste, permanent in air and readily soluble in water.

Sodium Chloride (Common Salt), $NaCl$ = 58·5, which h~s a little application in colour making, is sufficiently ₃ in the form in which it is generally used for house .ɔıα purposes.

SALTS OF THE ALKALINE EARTH METALS.

The metals which are known as the alkali earth metals have much similarity with the alkali metals in their ｡ ｢˙ pounds, with the difference that their alkalinity is ₁. .｡˙ less, and that their salts are much less soluble in water. , ·

colour making the compounds of three of these metals—calcium, barium and magnesium—are used.

Calcium Compounds.

The most important calcium compounds are lime and carbonate and phosphate of lime. Carbonate of lime is used as a pigment, and will be dealt with in detail among the mineral colours; it will be but briefly described here.

Calcium Oxide (Quicklime), CaO = 56.—When chalk and limestone, which consist of calcium carbonate, are heated, carbonic acid is evolved, and calcium oxide, commonly called quicklime, is left. For our purposes only very pure quicklime is to be used. Its ordinary impurities are iron oxide and magnesia: the former is found in lime made from red or brown limestone; the latter in lime made from dolomitic limestone. The presence of oxide of iron is recognised by the reddish tinge of the quicklime; if magnesia be present, a small quantity of the quicklime, when mixed with a very large quantity of water, leaves an insoluble residue which consists of magnesia.

When quicklime and water are brought together they unite very energetically and form calcium hydroxide or slaked lime. According to the quantity of water used for slaking, either dry slaked lime, lime paste, or milk of lime is produced, all of which find a use in colour making.

Calcium Hydroxide (Slaked Lime), $Ca(OH)_2$ = 74.—In order to prepare slaked lime, which contains lime united with just the necessary quantity of water, the pieces of quicklime are sprinkled with water from a watering can. The water is rapidly taken up and the sprinkling is repeated until the lumps begin to fall to a fine powder; in the process the lime becomes very hot. The slaked lime is then passed through a sieve in order to separate the larger pieces of quicklime which have not been slaked; the powder must be kept in well closed packages, since it energetically absorbs carbonic acid out of the air.

If so much water is added to the lime that a homogeneous wet mass is formed which can be readily moved with a shovel, one has then lime paste, which can be conveniently kept in pits as the masons do; it may be stored in this way for many months without appreciable alteration, still it is better to keep it covered. To prepare milk of lime, so much water is used in slaking that a milky liquid is formed, or the lime paste is mixed up in the proper quantity of water. Slaked lime dissolves in 700 to 800 parts of water; on standing, the undissolved slaked lime settles to the bottom of the milk of lime: thus it is better to prepare milk of lime immediately before use, and to stir it well to prevent the settling of the solid particles.

Slaked lime in one of its forms is often used instead of the more costly caustic soda in order to precipitate metallic oxides from their salts.

At times one finds a too strongly burnt lime, so-called "dead-burnt" lime, which is very slowly slaked by water. Such quicklime is slaked by allowing it to lie in water for days, or by means of hot water, which accomplishes the slaking more quickly.

Calcium Carbonate, $CaCO_3 = 100$, is found naturally in large quantities as chalk, which consists of the skeletons of extremely small animals. By powdering and levigating, it is converted into a soft powder, which is used to lighten the shade of lakes and other colours.

Calcium Sulphate (Gypsum), $CaSO_4.2H_2O = 172$. — This mineral, when finely powdered, is added to some colours.

Calcium Phosphate (Bone Ash), $Ca_3(PO_4)_2 = 310$, is sometimes used to lighten the shade of certain colours which might be injured by calcium carbonate. It comes in large quantities as a fine, white powder from South America. Naturally, only quite white bone ash can be used; if not completely burnt it is grey, and will then impair the shade of colours with which it is mixed.

Magnesium Carbonate (Magnesia), $MgCO_3 = 84$, is also used as an addition to colours in order to obtain pale shades. It is most cheaply obtained by dissolving magnesium sulphate (Epsom salts) in water and adding soda solution so long as a precipitate forms, which is then washed and dried. The magnesium carbonate prepared in this way is a very fine, light powder, insoluble in water, which can be mixed with the most delicate colours without harming them. White magnesia is an extremely light powder, which may be used when it can be bought at as low a price as the above preparation.

Barium Compounds.—The raw material used for the preparation of barium pigments is either barium sulphate (barytes, heavy spar) or barium carbonate (witherite). The latter is much more rare than barytes, which is almost exclusively employed in the preparation of barium compounds. The barium compounds of particular importance for our purpose are barium chloride and nitrate.

Barium Chloride, $BaCl_2 . 2H_2O = 244$.—This salt is now a common article of trade, and can be bought at a low price. When pure it forms colourless crystals readily soluble in water. If the colour maker is able to get cheap barytes and fuel, it may be advantageous for him to prepare barium chloride himself.

To prepare barium chloride from barytes, the latter is very finely ground and levigated, intimately mixed with coal, and the mixture subjected to a very high temperature, when barium sulphide is formed, which is dissolved by washing out the mass with water and converted into barium chloride by adding hydrochloric acid, sulphuretted hydrogen being evolved.

The best method is to mix 4 parts of barytes with 1 part of bituminous coal and so much coal tar that a plastic mass is formed, which is well kneaded and made into small cylinders 3 centimetres in diameter and 10 centimetres long. These cylinders are placed in layers in a cylindrical furnace with a good draught, which contains at the bottom a layer of coal 15 to 20 centimetres thick, then a layer of the cylinders, then again coal, and so on until the furnace is full. The lowest layer of coal is lighted, and the whole burnt at a bright red heat, when the barium sulphate is changed into sulphide. Hydrochloric acid is poured over the residue and the insoluble part, consisting chiefly of unaltered barytes, is used for the next operation.

Witherite (barium carbonate) can be converted into barium chloride in a very simple manner. Hydrochloric acid is added, in which it dissolves with the evolution of carbonic acid. The solution is allowed to stand twenty-four hours with excess of witherite; the whole of the dissolved iron is thus precipitated. The solution is then filtered, evaporated down and left, when pure barium chloride crystallises out.

Barium chloride and all soluble barium salts should only be dissolved in pure water (rain or distilled). Water which contains carbonates or sulphates always gives a turbid solution by precipitating barium carbonate or sulphate.

Aluminium Compounds.

The compounds of the earth metal aluminium play a very important part in colour making, since they form beautifully coloured compounds with many organic colouring matters. Formerly alum was the only material used in colour factories for the preparation of the alumina compounds; at present aluminium sulphate is used, and when it is sufficiently pure it is the most valuable material, because it contains the greatest quantity of alumina.

Under the designation of alum only one compound, the so-called potash alum, was at one time found in commerce, but now there are other alums, which contain soda or ammonia in place of potash. These salts are of equal use in colour making to potash alum. The preference is to be given to the compound which contains the largest proportion of alumina. The chief point to be observed in connection with alumina compounds for use in colour making is that they shall be free from iron, because iron oxide, which would be precipitated out of the solution along with the colours, in consequence of its red colour would spoil the shade of the pigment.

Aluminium Sulphate (Sulphate of Alumina), $Al_2(SO_4)_3 \cdot 18H_2O = 664$.—Any manufacturer who can obtain cheap china clay (kaolin) and sulphuric acid can himself prepare this compound with advantage. The apparatus used for this purpose is an iron pan containing sand, in which is placed a large earthenware dish. In this dish are put very finely-ground kaolin and strong sulphuric acid, and the mixture is heated so strongly that the acid boils, evolving heavy, white vapours. It is absolutely necessary to heat in this manner in order to avoid dangerous accidents. Sulphuric acid bumps so violently on boiling that it may even break a thick earthenware dish. The use of a sand bath makes the bumping harmless.

China clay, which consists of silicate of alumina, is decomposed by heating with sulphuric acid into silicic acid and sulphate of alumina. The original milky liquid becomes more transparent during boiling, and has at last the appearance of starch paste. Kaolin contains varying quantities of silica. The quantity of sulphuric acid necessary for its decomposition can only be found by trial. The quantities are chosen so that a small amount of kaolin remains undecomposed in order that the aluminium sulphate shall contain no free sulphuric acid.

When the decomposition is finished the pan is allowed to cool and the solid mass is brought into a vat filled with water, in which it is stirred until dissolved; then the liquid is left until the jelly-like mass of silicic acid has sunk to the bottom, when the clear solution of aluminium sulphate is drawn off and can at once be used.

If solid aluminium sulphate is required—and this is to be recommended when large quantities are to be prepared—the solution is evaporated in earthenware dishes until a portion solidifies when dropped on a cold plate. The molten aluminium sulphate is then cast in prismatic blocks, which are preserved in boxes. These blocks are of a pure white colour and very crystalline; they dissolve with difficulty in cold, but readily in hot water without residue. The solution has an acid taste, even when it contains no excess of sulphuric acid. When the blocks have a yellowish tinge, this denotes the presence of iron, and the solution must be freed from this impurity. Nowadays sulphate of alumina can be obtained so cheaply that it is hardly of advantage to make it.

The Alums.

These are double salts of aluminium sulphate and potassium, sodium or ammonium sulphate. There are also other double sulphates known as alums which, in place of aluminium, contain chromium, iron, etc., but they are not of interest here. It may still be observed that all alums, whatever their composition, possess the property of crystallising together from mixed solutions, so that crystals can be obtained in which every existing alum is contained.

The potassium, sodium and ammonium aluminium alums are used in colour making.

Potassium Aluminium Alum, $KAl(SO_4)_2.12H_2O = 474$.—This is the substance commonly called alum. Like all alums it crystallises in fine octahedral crystals, which at first are quite transparent, but slowly effloresce in the air and become covered by a white powder. It dissolves with difficulty in cold, but readily in hot water. The solution has at first a sweet taste, with an astringent after-taste.

Alum comes into the market in different forms, of which the following are the most important: as crystallised alum, in the form of large crystals united together; as alum meal, a coarse crystalline powder obtained by rapidly cooling and stirring a hot alum solution. On account of the larger surface this form dissolves more quickly than the large crystals. Roman alum is the name of a variety chiefly imported from Italy; it owes its reputation to its great purity—it contains a very small quantity of iron.

In order to prepare alum quite free from iron from the ordinary alum containing iron, it is recrystallised, that is, as much as possible is dissolved in boiling water and the solution quickly cooled with continual stirring; the small crystals so obtained are then washed with cold water. The residual saturated solution of alum, which contains the greater part of the iron, can be used for the preparation of those colours which are not injured by the presence of iron.

The solubility of alum in water varies greatly at different temperatures. The table gives the weight of alum dissolved by 100 parts of water at different temperatures.

Temperature.	Crystallised Alum.	Anhydrous Alum.
0° C.	3·90	2·10
10° ,,	9·52	4·99
20° ,,	15·13	7·74
30° ,,	22·01	10·94
40° ,,	30·92	14·88
50° ,,	44·11	20·09
60° ,,	66·65	26·76
70° ,,	90·67	35·11
80° ,,	134·47	45·66
90° ,,	209·31	58·64
100° ,,	357·48	74·53

When potash alum is heated it loses water, 75 per cent. of the total at 61° C.; at 92° C. it melts completely, and all the water is lost by continued heating at 100° C. The residue is known as burnt alum.

In alum the whole acidity of the sulphuric acid is not neutralised; the solution has always an acid reaction; if soda solution is added, the escaping carbonic acid causes the liquid to effervesce. If soda solution is added until a further addition would cause a precipitate, a solution of so-called neutral alum is formed which has no longer an acid reaction. Neutral alum is occasionally required in colour making. In preparing it the soda solution must be added with great care when the liquid is near its point of neutralisation. Any addition of soda solution after this point is reached will cause a separation of alumina. This is not desirable, since it is generally only wished to precipitate the alumina in combination with colouring matters.

Roman Alum.—Under this name, or that of "cubic alum," a variety of alum is sold, generally at a rather higher price than ordinary alum, from which it is distinguished by its crystalline form. Ordinary alum forms octahedral crystals often the size of a child's head, but cubic alum well formed cubes.

The property of crystallising in cubes may be imparted to any alum solution by the addition of a little potash. Much so-called Roman alum is made in German works in this way. When this alum contains very little iron it is quite equal in quality to the best Roman alum, for the higher value of the latter is entirely due to its small content of iron. The alum prepared in the province of Naples is still better than Roman alum; it contains less iron.

When alum is required for the preparation of lakes of bright and delicate shades, it is indispensable to use a preparation free from iron, because the brownish yellow oxide of iron would appreciably injure the shade. Alum, free from iron, is most simply prepared by dissolving alum in boiling water, running the boiling solution quickly through a cloth, and quickly cooling with constant stirring. The alum meal prepared in this way contains much less iron than the original alum, the iron salts remaining dissolved in the mother liquor. When this alum meal is collected and cold water poured over it to free it from mother liquor, it is generally sufficiently pure to be used for any purpose, but if not, it is again recrystallised. The alum liquors containing the iron are used for the preparation of colours which are not injured by the presence of iron.

To test alum for iron yellow prussiate of potash is used, which gives a blue precipitate with ferric salts. The test is carried out by dissolving 10 grammes of alum in 1 litre of water, placing the solution in a tall, narrow cylinder standing on white paper, adding 10 to 20 drops of a saturated solution of yellow prussiate, and well stirring. On looking down through the liquid, if a distinct colouration is at once evident, the alum contains much iron, and must be recrystallised; indeed, the crystals would generally be coloured yellow. On the contrary, if the solution does not show a blue tint until after standing several days, the alum contains but a small quantity of iron, and can be used for most purposes without further purification. Alum quite free from iron is a rare commercial article; the test will generally show a feeble blue colouration. If this is not intense and no blue precipitate is deposited at the bottom, the alum is tolerably pure, and can be used in colour works. The longer the time before the blue colouration appears the poorer is the alum in iron.

Soda Alum, $NaAl(SO_4)_2 \cdot 12H_2O = 458$, is made in some alum works. It has the greatest similarity in properties with the potash salt, but is distinguished by a much greater solubility in water and more rapid efflorescence in air. Soda alum can be bought at varying prices; that containing iron is much cheaper than that free from iron. When the latter is to be bought at a fair price it is to be preferred to potash alum, since, as we shall show later, it contains a larger proportion of alumina.

Ammonia Alum, $(NH_4)Al(SO_4)_2 . 12H_2O = 453.$—This compound of aluminium sulphate and ammonium sulphate is now often met with; the more expensive potassium sulphate in ordinary alum is replaced by ammonium sulphate, which is cheaply obtained from the ammoniacal liquor of the gas works.

Ammonia alum is better for our purpose than potash alum since it contains more alumina, is generally cheaper and dissolves more easily in water. Unfortunately most commercial ammonia alum contains so much iron that it has to be recrystallised before it can be used in colour works.

One hundred parts of water dissolve at different temperatures the quantities of ammonia alum given in the table:—

Temperature.	Crystallised Ammonia Alum.	Anhydrous Ammonia Alum.
0° C.	5·22	2·62
10° ,,	9·16	4·50
20° ,,	13·66	6·57
30° ,,	19·29	9·05
40° ,,	27·27	12·35
50° ,,	36·51	15·90
60° ,,	51·29	21·95
70° ,,	71·97	26·09
80° ,,	103·08	35·19
90° ,,	187·82	50·30
100° ,,	421·90	70·83

Of the different alums, ammonia alum contains the largest and potash alum the smallest proportion of alumina. The composition of the three commonly occurring alums is given in the following table:—

	Potash Alum.	Soda Alum.	Ammonia Alum.
Potash, K_2O	9·95	—	—
Soda, Na_2O	—	6·80	—
Ammonia, NH_3 . . .	—	—	3·89
Alumina, Al_2O_3 . . .	10·83	11·20	11·90
Sulphuric acid, SO_3 . .	33·71	34·90	36·10
Water, H_2O_4	45·51	47·10	48·11

Thus ammonia alum is to be preferred to soda alum and soda to potash alum, whilst the latter is used on account of its greater purity. In a colour works, in which large quantities of alum are used, it is advantageous to work with ammonia alum which is recrystallised on the works.

The alums and aluminium sulphate are the alumina compounds in ordinary use in colour works; aluminium acetate could also be used if it were to be had at a reasonable price.

Alumina, $Al_2O_3 = 102$, and Hydrate of Alumina.—Pure alumina, or rather hydrate of alumina, is required in the preparation of many colours. When the solution of an aluminium salt is precipitated by soda, the carbonic acid escapes with effervescence, and a gelatinous precipitate is formed which it is extremely difficult to wash clean. The precipitate, which consists of hydrate of alumina, shrinks very greatly in drying, and turns to a horny mass; when strongly heated it loses water, and becomes a white, insoluble powder of anhydrous alumina.

A variety of hydrate of alumina, heavy, and therefore easily washed, is obtained by boiling a solution of alum with a plate of zinc lying on a copper plate until all the alumina has separated. By collecting this on a filter and pouring hot water over it a number of times it is obtained quite pure.

Alumina plays a particularly important part in the manufacture of cobalt colours. When treated with a cobalt salt and heated it takes a fine blue shade.

In the foregoing those metals and their compounds have been treated which are extensively used in preparing colours without themselves forming coloured compounds. The chromates and prussiates are an exception to this. The ammonia compounds, the alkalis and alkaline earths, also the acids, are used in making many colours, although they do not contain colouring principles. The alumina compounds are in similar case. Themselves colourless, they form at the same time a carrier for the coloured compound and bring it into a suitable form for use as a pigment.

An example will explain what we mean by "carrier" of the colouring matter. Logwood contains a very handsome colouring matter which can be extracted by water. In order to be able to employ this colouring matter as a pigment it is combined with alumina, a compound insoluble in water being formed, which is called a lake. In this compound the alumina is to be regarded as the carrier of the colouring matter, which it has fixed in an insoluble form.

In dyeing, which in many respects is closely allied to colour making, the property of certain metallic compounds, themselves colourless, of fixing dyes is commonly utilised, the metallic compound being called the mordant. The fabric is first prepared with the metallic compound or mordant, and the colour then formed by bringing the mordanted material in contact with the colouring matter.

The "heavy metals" form, among their numerous compounds, a great number of coloured substances, and several of them are distinguished by a great wealth of coloured derivatives; for example, copper, chromium and cobalt form coloured compounds only. Although the use of pigments derived from the heavy metals has been considerably restricted in recent years by the discovery of a series of colouring matters which replace them, yet they are now, and always will be, of very great importance in the manufacture of colours. It is, therefore, necessary briefly to describe the various metals which are used in colour making, so that the manufacturer may know what metals produce harmless colours, permanent and unaltered by the atmosphere, and what do not.

The metals are divided into two great groups, designated, according to their specific gravities, the group of the light and of the heavy metals. The light metals comprise the alkali, alkaline earth and earth metals, whose important compounds we have just described. The specific gravity of each of these metals is less than five times that of water.

The heavy metals have a specific gravity exceeding 5; they are generally divided into groups, which are known by the name of the commonest metal in the group. These are as follows:—

Zinc group	Zinc, $Zn = 65$.
Iron group	Iron, $Fe = 56$.
Tungsten group	Tungsten, $W = 184$.
Antimony group	Antimony, $Sb = 120$.
Tin group	Tin, $Sn = 119$.
Lead group	Lead, $Pb = 207$.
Silver group	Silver, $Ag = 108$.
Gold group	Gold, $Au = 197$.
Platinum group	Platinum, $Pt = 194$.

To the zinc group belong the metals zinc, cadmium and indium, of which only the two first are of importance here. The iron group comprises iron, manganese, cobalt, nickel, chromium and uranium, all of which are used in the manufacture of colours The antimony group contains antimony and bismuth, the latter of which is of little importance. To the tin group belong tin and the rare metals titanium, zirconium, thorium, niobium and tantalum; of these tin alone is important in colour making. In the lead group are lead and thallium; lead produces many important pigments. The silver group contains silver, copper and mercury, the two latter of which are important. Of the gold group gold alone is of interest, and its importance has been diminished by the discovery of far cheaper substances which replace it. In the platinum group, which contains platinum, iridium, rhodium, ruthenium, palladium and osmium, only platinum itself is used as a colour; it is employed in porcelain painting to produce the peculiar metallic shimmer known technically as "lustre."

The behaviour of the compounds of the metals mentioned above towards sulphuretted hydrogen is of the greatest importance to the colour maker, since on it depends the alterability of the pigments when exposed to the atmosphere. Many metallic compounds are unaltered by the sulphuretted hydrogen present in the air, whilst others are in a high degree affected by it and become gradually darker, so that their colour may in the end approximate to black.

The pigments containing lead, copper, mercury and bismuth are extremely susceptible to the action of sulphuretted hydrogen, by the action of which they form black compounds. Since colours which contain these metals are not permanent, but darken considerably, endeavours have been made for a long time to replace them by others not susceptible to the action of sulphuretted hydrogen. Thus it is desirable to manufacture only colours free from metals forming black compounds with sulphuretted hydrogen. For the same reason pigments which contain sulphur should not be mixed with those containing metals which form black sulphur compounds.

The rules laid down in the preceding paragraphs are of the greatest importance for the artist, for by following them he will succeed in composing a "permanent palette," that is, containing only such colours as will not, by their composition, bring about the speedy ruin of the painting.

Although the majority of the compounds of the heavy metals are poisonous, some possess this property in an eminent degree. These are chiefly colours which contain arsenic, antimony, copper and lead. So far as it is possible, these colours should be dispensed with and harmless pigments sold in their place, though this is not always possible, since several poisonous colours cannot be replaced by innocuous ones.

Compounds of the Heavy Metals.

Zinc Compounds.—Zinc oxide, ZnO, and zinc sulphate, $ZnSO_4.7H_2O$, are the compounds of this metal used in the colour industry. Zinc oxide, which is used as a white pigment, is a powder which turns yellow when heated, and is not acted upon by sulphuretted hydrogen. Zinc sulphate (white vitriol) occurs as colourless crystals, or more frequently as greyish white crystalline masses. The freedom of zinc sulphate from iron is of particular importance; the commercial article is seldom satisfactory in this respect. In order to free commercial zinc sulphate from iron, the property of zinc hydroxide of precipitating iron oxide from neutral solutions may be employed. The zinc sulphate is dissolved in water, and ammonia added in small quantities until the precipitate of zinc hydroxide remains on stirring. When the liquid is left in contact with the precipitate and stirred up once or twice a day, if iron is present the precipitate will turn yellowish brown, owing to the separation of ferric hydroxide, and in the course of a few days all the iron will be removed from solution. The liquid should then give no blue colouration with yellow prussiate of potash.

Zinc oxide is used as a white pigment, zinc chromate as a yellow, and zinc cobalt compounds as green colours.

Cadmium Compounds.—Cadmium is a metal which possesses great similarity to zinc, with which it occurs in nature. In the preparation of cadmium compounds the metal is generally used. This is dissolved in dilute sulphuric acid, hydrogen is evolved, and a solution of cadmium sulphate obtained.

Cadmium is used in colour making only for the preparation of the beautiful cadmium yellows. .

Iron Compounds.—These are of the greatest importance to the colour maker. Several, in which iron alone is the colour principle, are very valuable: ochre, rouge, Venetian red, sienna and umber, for example. Iron compounds are also used in the preparation of many colours. The most important is:—

Ferrous Sulphate (Green Vitriol, Copperas), $FeSO_4.7H_2O$.—This substance, which occurs commercially in a form of great purity at a very low price, is generally the starting-point in the preparation of iron pigments. When pure, it forms fine sea-green crystals, with an astringent metallic taste, which are not poisonous and are easily soluble in water. After long exposure to the air, ferrous sulphate becomes covered with an ochre-coloured crust, consisting of basic ferric sulphate. The ferrous oxide contained in the green vitriol has united with oxygen and been converted into ferric oxide. The latter requires a larger quantity of acid than ferrous oxide for the formation of soluble salts, so that an insoluble basic salt is separated. The same thing occurs when a solution of ferrous sulphate is exposed to the air.

When green vitriol, or any other ferrous salt, is exposed to the action of oxidising agents, such as chlorine or nitric acid, the iron is rapidly changed into the ferric state. This transformation is of particular importance in the manufacture of certain blue pigments.

We give in a table the relation between the percentage

of crystallised ferrous sulphate ($FeSO_4 \cdot 7H_2O$) contained in a solution at 15° C. and its specific gravity :—

Specific Gravity.	Percentage of Ferrous Sulphate.	Specific Gravity.	Percentage of Ferrous Sulphate.
1·000	0	1·118	21
1·005	1	1·125	22
1·011	2	1·131	23
1·016	3	1·137	24
1·021	4	1·143	25
1·027	5	1·149	26
1·032	6	1·155	27
1·037	7	1·161	28
1·043	8	1·168	29
1·048	9	1·174	30
1·054	10	1·180	31
1·059	11	1·187	32
1·065	12	1·193	33
1·071	13	1·200	34
1·077	14	1·206	35
1·082	15	1·213	36
1·088	16	1·219	37
1·094	17	1·226	38
1·100	18	1·232	39
1·106	19	1·239	40
1·112	20	—	—

Yellow and red prussiate, which have been already mentioned, also belong to the iron compounds. They have been separately mentioned because the iron is contained in them in a peculiar form as a portion of an organic radical.

Ferrous Chloride, $FeCl_2$, may sometimes be used instead of green vitriol. When iron is dissolved in hydrochloric acid hydrogen is given off and a solution of ferrous chloride is obtained; but when rouge is dissolved in the same acid, ferric chloride is formed. When iron is dissolved in nitric acid, in consequence of the oxidising properties of this acid a ferric salt is obtained. Iron forms two series of salts: in the ferrous compounds the iron is in the same form as in green vitriol and the corresponding salts; in the ferric compounds the iron is contained in a higher state of oxidation. By powerful oxidising agents, as nitric acid or chlorine, ferrous compounds are converted into ferric.

Manganese Compounds.—Manganese (Mn) is a metal whose compounds show great similarity with those of iron, like which it forms two oxides (also others), manganous oxide (MnO) and manganic oxide (Mn_2O_3). The salts of manganous oxide are not oxidised in the air like those of ferrous oxide.

The raw material used in the preparation of manganese compounds is the mineral pyrolusite, which is manganese dioxide (MnO_2).

Manganese sulphate ($MnSO_4$) forms rose-red crystals containing varying quantities of water. The residues from the preparation of chlorine can be used as the material for the preparation of colours. According as pyrolusite and hydrochloric acid or pyrolusite, salt and sulphuric acid are used for this purpose a solution of manganous chloride or sulphate is obtained.

Manganese compounds have but a restricted use in colour making.

Nickel Compounds are generally coloured green, but they are not used as pigments.

Cobalt Compounds.—Among these are found many important pigments. All cobalt compounds are coloured; in beauty and variety of shade they can only be compared with those of chromium. In properties cobalt is very similar to iron and nickel.

The form in which cobalt is used in preparing colours is either cobalt nitrate, $Co(NO_3)_2.6H_2O$, or cobalt chloride, $CoCl_2.6H_2O$. Both salts are articles of commerce, but generally they are so dear that it is more profitable for the colour maker to prepare them direct from the cobalt minerals. A simple method for preparing cobalt compounds from the ores is therefore given. The most important cobalt ores are speiss cobalt, a compound of cobalt and arsenic, and cobalt glance, a compound of cobalt, arsenic and sulphur. The former mineral often contains only small quantities of cobalt, and it is advisable for our purposes to use cobalt glance, which contains from thirty to forty per cent. of cobalt. This mineral is first roasted, that is, is heated with a plentiful air supply, by which means the arsenic is driven off. On account of the poisonous nature of the arsenic vapours the roasting must be conducted in a furnace with a very good draught.

Under the name of zaffre, roasted cobalt ores come into commerce. These may be used in the preparation of cobalt compounds, by which means the operation of roasting is avoided. According to the quality of the ore which has been used to obtain zaffre, it contains a very varying proportion of cobalt. The varieties richer in cobalt must be used; they are technically known by the mark FS, or FFS (the best).

The roasted cobalt ores or zaffre are treated with fused acid potassium sulphate, when the salts of iron and manganese are decomposed, whilst cobalt and nickel sulphates remain unchanged. In a Hessian crucible are melted 300 parts of acid potassium sulphate, and 100 parts of the powdered zaffre are gradually added, mixed with one part of green vitriol and one part of saltpetre; the mixture is heated so long as sulphuric acid escapes. The mass is then boiled with water and the red solution treated with sulphuretted hydrogen so long as a precipitate is formed; this may contain copper, manganese, and bismuth. After filtering, soda is added to the boiling liquid; cobalt carbonate is precipitated, which can be converted into nitrate or chloride by solution in the corresponding acid. If cobalt sulphate is required, the solution, after treatment with sulphuretted hydrogen, need only be evaporated to crystallisation, when the sulphate separates in fine red crystals. Cobalt nitrate and chloride are very soluble in water; to obtain them their solutions must be strongly evaporated and quickly cooled whilst stirring. The crystals of cobalt nitrate and chloride absorb moisture from the air and deliquesce; they must be kept in glass vessels with well-ground stoppers.

The cobalt compounds which are to be used in colour making must be free from iron, nickel and arsenic, which would detract from the cleanness of the colours. If the precipitate produced by soda contains iron it is mixed with excess of solution of oxalic acid, and after a few hours the cobalt oxalate is filtered from the liquid, in which all the iron is dissolved. The cobalt oxalate can then be converted into nitrate or chloride by treatment with nitric or hydrochloric acids.

These salts form the material for the preparation of the cobalt compounds, a large number of which are used as extremely durable red, blue and green pigments; several of them, such as cobalt blue, cannot be exactly replaced by other pigments. On account of the industrial importance of the cobalt colours, these directions for the preparation of the soluble cobalt salts from the ores have been given with some detail. The preparation of the cobalt colours will be given *in extenso* later on.

Chromium Compounds. — As the name indicates, this metal yields numerous coloured compounds ($\chi\rho\tilde{\omega}\mu\alpha$, colour); in fact, only coloured chromium compounds are known, and the colours are most varied—yellow, green, red and violet. On this account the chromium compounds are among the most important used in colour making; a great number of colours are prepared by their aid. Chrome ironstone, as we have already stated, is the raw material for the preparation of chromium compounds. From it potassium bichromate is made on a large scale in special works, so that no colour maker is compelled to prepare chromium salts himself.

When the chromium pigments contain no metal blackened by sulphuretted hydrogen, they have the desirable property of being unaltered by the atmosphere. Like the cobalt compounds, they are distinguished by their great stability when heated; on this account, they have a large use in porcelain painting.

Molybdenum, Tungsten and Vanadium Compounds on account of their cost have a very limited use as pigments. Molybdenum compounds are obtained from molybdic acid; compounds of tungsten from the metal; and those of vanadium from ammonium vanadate.

Antimony Compounds can be used in the preparation of several pigments, but, on account of their behaviour towards sulphuretted hydrogen, the pigments cannot be regarded as really permanent, and their use is generally diminishing. The so-called antimony vermilion is the only antimony compound at all extensively employed.

Bismuth Compounds possess properties very similar to those of antimony. Only one bismuth preparation is used as a pigment, and this is very sensitive to the action of sulphuretted hydrogen, being changed into black bismuth sulphide.

Tin Compounds are employed in two ways: some are themselves colours, such as stannic sulphide (mosaic gold); others, themselves colourless, are used in making pigments, as stannous and stannic chlorides.

Stannous Chloride, $SnCl_2.2H_2O$, is obtained when tin is dissolved in hydrochloric acid, hydrogen being evolved. Stannic Chloride, $SnCl_4$, is formed when tin is dissolved in a mixture of hydrochloric and nitric acids (*aqua regia*).

Tin compounds have, similarly to aluminium salts, the property of forming coloured insoluble compounds (lakes) with many organic colouring matters. Their use for this purpose is extensive.

Arsenic Compounds formerly played an important part

in colour making. They were used in the manufacture of a large number of pigments, very beautiful but extremely poisonous. At the present time, we can, fortunately, entirely dispense with arsenic in colour making; the arsenic colours can be replaced by others equally handsome and less, or not at all poisonous. The most important of the arsenic compounds is the trioxide As_2O_3, or Arsenious Acid, commercially known as white arsenic. This substance is obtained in large quantities as a by-product in the extraction of several metals. It forms masses which are either glassy or have the appearance of porcelain. Freshly sublimed arsenic trioxide is glassy; this form gradually changes into the porcellaneous variety, it dissolves with difficulty in water. A strong solution can only be obtained by boiling for many hours.

The compounds of arsenic with sulphur, formerly extensively used as pigments, have now almost fallen into disuse.

Lead Compounds belong to the substances most largely used in making colours. Unfortunately all lead colours have two very important drawbacks. They are all very poisonous and at the same time extremely sensitive to sulphuretted hydrogen, so that they are very considerably altered by the action of the small quantities of that gas contained in the atmosphere of an ordinary dwelling. A striking example of this is seen in the lead paints used on the doors of water-closets. The paint, at first pure white, becomes gradually darker, and at last almost black, the lead compound having been changed into black lead sulphide.

On account of this great sensitiveness of lead compounds, it would be better if they could be excluded from the list of colours. Great care must be taken not to mix lead compounds with others which contain sulphur; a discolouration of the mixture would be the inevitable result in a very short time.

The oxides of lead and a number of its salts are themselves pigments, for example, litharge, PbO, red lead, Pb_3O_4, and white lead (basic lead carbonate). These pigments are prepared on a large scale in particular works. At this point only those lead compounds will be mentioned which are generally used for the preparation of other lead pigments; they are: lead sulphate, nitrate, acetate and chloride.

Lead Sulphate, $PbSO_4$, is formed when sulphuric acid or the solution of a sulphate is added to the solution of a lead salt; so obtained it is a white crystalline powder insoluble in water. This substance is generally not made in colour works, but is purchased from chemical works or dye houses, of which it is a by-product. In this form (lead bottoms) it is generally not sufficiently pure, but contains admixtures of sulphuric acid or aluminium salts, from which it is freed by washing. The lead sulphate is stirred up in water, the heavy precipitate allowed to settle, the wash water drawn off, and after repeating this process until the wash water no longer shows an acid reaction, the purified precipitate is dried. In this condition it is a heavy, white powder, and can alone be ground into paint. But on account of its crystalline nature, which reduces the covering power, such use is inadvisable.

Lead Nitrate, $Pb(NO_3)_2$. — This very important compound may be bought, but it is advisable to prepare it in the works. Water is placed in a wooden tub, then half the volume of nitric acid is added, and finely powdered litharge gradually stirred in, the liquid being kept in constant movement. When it is seen that the litharge is only slowly dissolved, the liquid is well stirred after each addition of litharge and then tested by litmus paper. When this is no longer reddened, the nitric acid is completely saturated, and the liquid contains only lead nitrate in solution. It is allowed to stand until the insoluble portions have settled, and then drawn off into another vessel where crystals of lead nitrate separate in a few days. If the salt be required in solid form, the solution may be evaporated in earthenware dishes; generally the solution is used as it is obtained.

Pure lead nitrate forms white crystals which are not particularly soluble in water, 1 part requiring 2 parts of water at the ordinary temperature. Lead nitrate is decomposed on heating, like all nitrates, and litharge remains. The solution of this salt is used in the preparation of those lead pigments which are obtained by precipitation, for example, chrome yellow.

Lead Acetate, $Pb(C_2H_3O_2)_2.3H_2O$. — The compounds of lead with acetic acid are of great importance. Two of these are to be considered: neutral lead acetate, commonly known as sugar of lead, and basic lead acetate. It may be advisable to manufacture both these compounds, the latter always. Neutral lead acetate comes into commerce in the form of colourless heavy crystals, which are often covered by a white powder of the basic acetate; they dissolve readily in water, and the solution has a sweetish taste, hence the name "sugar of lead". Frequently the solution is very turbid; this is caused by the carbonates contained in the water. The turbidity may be removed by the addition of a little acetic acid. It is only economical for the colour maker to prepare sugar of lead when he can obtain cheap raw materials, lead or litharge and vinegar. Pyroligneous acid may also be used if it is colourless, its odour being without importance for this purpose.

The best method for preparing lead acetate from litharge is to place the vinegar in a tub and hang in it a strong linen bag filled with finely-ground litharge. The tub is kept covered for a few days and its contents then tested with red litmus paper. When this is turned blue, the liquid is drawn off and vinegar gradually added whilst stirring, until blue litmus paper is just turned red. In this process, after neutral lead acetate has been formed, more lead oxide is dissolved, and the liquid thus acquires an alkaline reaction. The further addition of acetic acid reconverts the basic salt into neutral lead acetate.

Lead acetate solution may, with advantage, be prepared directly from metallic lead. For this purpose, lead is granulated by melting and pouring in a thin stream into cold water, where it solidifies in irregular pieces. This is done in order to give the lead as large a surface as possible. Three high narrow tubs placed one above the other so that liquid may flow from the highest to the middle, and from this into the lowest, are filled with granulated lead. Vinegar is placed in the uppermost vessel to cover the lead; after twenty-four hours it is allowed to flow into the middle, and after a further twenty-four hours into the lowest tub. In this way, a solution of basic lead acetate is formed, to which the necessary quantity of acetic acid is added to bring it into the neutral condition. If crystalline lead acetate is required, the liquid is evaporated down and quickly cooled with stirring, so that small crystals are formed. Generally, however, evaporation is unnecessary, since lead acetate is always used in solution in preparing colours.

If the lead acetate solution be not colourless, which is generally the case when coloured acetic acid is used, the defect may be removed by stirring a little bone black into the liquid and filtering after twenty-four hours, when a completely colourless solution is obtained.

It is always necessary to know exactly how much lead acetate is contained in the solutions prepared by these processes. The lead or the litharge is therefore weighed and the volume of the lead acetate solution measured. One hundred parts by weight of crystallised lead acetate are obtained from 62·54 parts of lead.

Lead acetate solutions must be kept in well-covered

vessels. The carbonic acid of the air will turn the liquid turbid. The turbidity may be removed by the addition of acetic acid.

In the following table is given the percentage of crystallised lead acetate contained in solutions of different specific gravities:—

Specific Gravity.	Crystallised Lead Acetate per cent.	Specific Gravity.	Crystallised Lead Acetate per cent.	Specific Gravity.	Crystallised Lead Acetate per cent.
1·0000	0	1·1159	17	1·2578	34
1·0064	1	1·1234	18	1·2669	35
1·0127	2	1·1309	19	1·2768	36
1·0191	3	1·1384	20	1·2867	37
1·0255	4	1·1464	21	1·2966	38
1·0319	5	1·1544	22	1·3064	39
1·0386	6	1·1624	23	1·3163	40
1·0453	7	1·1704	24	1·3269	41
1·0520	8	1·1784	25	1·3376	42
1·0587	9	1·1869	26	1·3482	43
1·0654	10	1·1955	27	1·3588	44
1·0725	11	1·2040	28	1·3695	45
1·0796	12	1·2126	29	1·3810	46
1·0867	13	1·2211	30	1·3925	47
1·0939	14	1·2303	31	1·4041	48
1·1010	15	1·2395	32	1·4156	49
1·1084	16	1·2486	33	1·4271	50

Basic Lead Acetate, $Pb(C_2H_3O_2)_2.2PbO$. — This salt may be regarded as a compound of neutral lead acetate with lead oxide. It is obtained by digesting vinegar with excess of litharge or with metallic lead, and also by treating lead acetate solution with litharge so long as the latter is dissolved. In the method last given, 100 parts of sugar of lead require about 118 parts of litharge to produce a saturated solution of basic acetate. The solution of this compound is alkaline; it turns red litmus paper blue. When exposed to the air, a turbidity is quickly produced owing to the separation of lead carbonate. In one white lead process basic lead acetate is the starting-point of the manufacture.

Lead Chloride, $PbCl_2$, is seldom used in making colours. It may be prepared by stirring powdered litharge in common salt solution until the powder appears white. This, when washed, constitutes basic lead chloride. On adding hydrochloric acid to the washed mass until the liquid remains acid, lead chloride is obtained in the form of crystalline needles, which are very little soluble in cold, but more easily in hot, water.

Like any other soluble lead salt, lead chloride may be used in the precipitation of colours, but is seldom employed on account of its small solubility. Basic lead chloride was, at one time, used as a white pigment, and after melting, by which it is turned yellow, as a yellow pigment; it is no longer in use for these purposes.

Copper Compounds.—These are generally green or blue, and have an extended use in the production of colours. The metallic copper which is used in the preparation of colours is of the ordinary commercial quality. The impurities which it contains are generally so small in quantity that they are without importance for our purpose.

Copper Sulphate (Bluestone, Blue Vitriol), $CuSO_4.5H_2O$.—This is the commonest of the commercial copper salts, and on that account deserves our especial attention. It forms large sky-blue crystals, which effloresce slightly in the air, possess an unpleasant metallic taste, and are poisonous, like all soluble copper compounds.

Copper sulphate comes into commerce in a very pure form, but some qualities contain zinc sulphate or ferrous sulphate. The presence of zinc may be detected most easily by boiling the solution with excess of caustic soda, when copper oxide separates as a black powder, whilst zinc oxide remains dissolved. When sulphuretted hydrogen is passed through the liquid, a white precipitate of zinc sulphide is formed.

Iron is detected by passing sulphuretted hydrogen through the solution so long as a precipitate is formed, allowing the liquid to stand in a covered vessel, pouring it off from the precipitate, adding nitric acid, boiling and adding a solution of potassium ferrocyanide; a blue precipitate denotes the presence of iron.

Copper sulphate is rarely found which is quite free from iron and zinc. If these impurities are present in but small quantity, the zinc not exceeding 1 per cent. and the iron at most 0·5 per cent., the copper sulphate may be regarded as sufficiently pure for our purposes.

It may be here remarked that copper sulphate obtained from mints is generally of great purity, and hence particularly adapted for colour making.

When copper sulphate and other copper salts are dissolved in water, pale blue flocks of copper carbonate generally separate. This is due to the carbonate of lime contained in the water. An addition of a few drops of sulphuric, nitric or hydrochloric acid suffices to prevent this separation.

Copper Nitrate, $Cu(NO_3)_2 . 6H_2O$.—This salt may occasionally be obtained in colour works as a by-product. When nitric acid is poured over copper, there follows a copious evolution of nitric oxide, which produces brown fumes of nitrogen peroxide in air. Nitric oxide may be used to convert ferrous into ferric salts, a transformation required in making Prussian blue. In working in this way, nitric acid is poured over copper contained in a vessel provided with a delivery tube for the gas. The blue solution is at once used. Pure copper nitrate forms fine blue crystals, which very readily deliquesce in the air. The solution is therefore generally used as it is prepared.

Copper Acetate, $Cu(C_2H_3O_2)_2 . H_2O$. — Copper is readily attacked by acetic acid. A number of salts are formed, of which some are used as pigments. For our purpose it will be sufficient to describe the manufacture of verdigris; few colour makers prepare any other copper acetate. A solution of this salt is most simply prepared in the following manner: Slaked lime is stirred with strong vinegar, and the solution left in contact with the excess of lime so long as it has a weak acid reaction. The solution, which contains acetate of lime, is poured into a solution of copper sulphate so long as a precipitate of sulphate of lime is formed. When this has been separated from the liquid, the latter is ready for further treatment. It contains only a very small quantity of dissolved sulphate of lime, which is not harmful in the preparation of colours.

In addition to the copper compounds mentioned here, several others were formerly used as pigments, or in the preparation of pigments which are no longer employed, because copper compounds of good colour can be obtained in a cheaper manner.

The same precautions should be taken in the use of copper colours which were mentioned for lead pigments; copper compounds are equally sensitive towards sulphuretted hydrogen, by which they are gradually discoloured.

Mercury Compounds.—Mercury forms compounds which, vermilion in particular, are used as pigments, and others which are used in the preparation of pigments. In many cases metallic mercury is the starting point in the preparation of the mercury compounds. The compounds commonly known as calomel and corrosive sublimate are also used.

Mercurous Nitrate, $HgNO_3$.—Nitric acid acts upon mercury in a manner differing according to its strength, and according to whether the mercury or the nitric acid is used in excess. In order to prepare mercurous nitrate, acid free from chlorine must be diluted at least with four times its volume of water, and the mercury must be in excess. On warming, the mercury is gradually dissolved, and, on cooling, the solution deposits colourless crystalline needles of the salt.

A further crop of crystals is obtained after evaporating the solution.

When the action of nitric acid is over, the solution must be at once separated from the excess of mercury to prevent the formation of basic salts. If the salt has been properly made, it is completely soluble in water, but if a lemon yellow precipitate is formed on dissolving, the nitrate contains a basic salt, which can only be dissolved by warming and adding more nitric acid.

Mercuric Nitrate, $Hg(NO_3)_2$, is most simply obtained by warming mercury with very strong nitric acid. The heating must be continued until a test portion of the solution no longer gives a precipitate with hydrochloric acid. When this solution is evaporated, nitric acid is given off, and a salt crystallising in white needles is obtained, which dissolves in water with the separation of a yellow basic salt. It is therefore better to use the hot solution, which contains a little free acid, without evaporating.

Instead of mercurous and mercuric nitrates, the corresponding sulphates may be used, but the chlorides are more frequently employed since they can be readily obtained from the makers.

Mercurous Chloride (Calomel), $HgCl$, is obtained pure by adding common salt solution to a solution of mercurous nitrate and washing the precipitate, which is insoluble in water.

Mercuric Chloride (Corrosive Sublimate), $HgCl_2$, a common article of commerce, is prepared by heating a carefully made mixture of mercuric sulphate and common salt, when mercuric chloride sublimes. It is a white crystalline mass, soluble in 13·5 parts of water at 20° C., and soluble in 3 parts of alcohol. Although all mercury compounds are very poisonous, corrosive sublimate requires particular care in handling, since its easy solubility makes it surpass all other mercury compounds in poisonousness.

The mercuric sulphate required in the above preparation is obtained by heating mercury with sulphuric acid. Corrosive sublimate can also be prepared by adding hydrochloric acid to mercurous nitrate, and heating, with gradual addition of hydrochloric acid until a clear solution is formed, from which mercuric chloride crystallises on cooling.

Silver Compounds.—Silver nitrate, $AgNO_3$, is the only one of importance here. It is obtained by dissolving silver in nitric acid, when a blue solution is obtained because commercial silver contains copper, evaporating the solution to dryness, melting the residue, and keeping it molten until all the copper nitrate is decomposed. This point is recognised when a small portion of the melt dissolved in water does not give a blue colouration with excess of ammonia. Fused silver nitrate forms a white crystalline mass readily soluble in water and turning black when exposed to light, like many other silver compounds.

Gold Compounds.—Gold is now very little used in preparing colours. The compound used for this purpose is gold chloride, $AuCl_3$, which is obtained by heating gold with hydrochloric acid, and adding nitric acid in small quantities until all the metal is dissolved. By careful evaporation of the yellow solution gold chloride is obtained in brownish yellow crystals, which easily dissolve in water.

The compounds of molybdenum, vanadium and uranium are less used than those of gold, yet these metals find a special use in the preparation of colours for porcelain painting and for colouring glass, for which they are of great importance. The preparation of their compounds from the raw materials is complicated and not remunerative to the colour maker; they should be obtained from chemical works.

In the foregoing, the most important compounds of inorganic origin used in making colours have been briefly described, less in order to teach the methods for their preparation than to give the manufacturer the means of learning their properties. Since the great development of chemical industries during the last decades it is more advantageous for the colour maker in most cases to draw his supply of these substances from works of good reputation than to make them himself; only in the case of a few substances, which are sold at unreasonably high prices, will it be profitable for him to prepare them himself.

CHAPTER VI.

THE MANUFACTURE OF MINERAL PIGMENTS.

By mineral pigments we understand those which consist of compounds of metals with elements such as sulphur, chlorine and iodine, or with compound radicals, such as cyanogen, or of metallic salts.

Looking at the classification of mineral pigments from the chemical standpoint, a grouping according to the constituents would appear preferable, and we should have groups of colours consisting of metallic oxides, sulphides, salts, etc. In such a division of pigments, according to their constituents, no regard would be taken of the colour of the pigment. The white zinc oxide would be placed in the same group with yellow lead oxide and red lead oxide; and red mercury sulphide and yellow cadmium sulphide would fall into the same group of the sulphur compounds.

The chemical composition of a pigment is of less importance to the colour manufacturer than its shade, and it therefore appears to us more reasonable to prefer to classify mineral pigments according to similarity of colour. We shall therefore place all those mineral pigments together which possess the same colour.

Common usage differs from the scientific in the description of colours. In the physical sense, yellow, red and blue are the so-called "simple colours," between which lie orange, green and violet as "mixed colours". Physics knows no white colour and no black colour, but describes white as a mixture of all the simple colours and black as the absence of colour. A grey or brown shade, produced by different mixtures of simple colours, is just as little known in the scale of colours as white or black.

The colour maker follows, as we have said above, the common manner of speaking; to him white and black are equally as much colours as red and green. Besides the pure principal colours (yellow, red and blue), and the mixed colours obtained from them (orange, green and violet), colour makers distinguish many shades of each colour — lemon yellow, sulphur yellow, cherry red, blood red, violet blue, etc. For the present purpose it is of great importance to accurately distinguish the several shades, for the value of many colours is in proportion to their beauty of shade. The colour maker is often required to produce a colour of some particular shade, which he accomplishes in many cases by a suitable alteration in the process by which the colour is made, in other cases by mixing different colours, in which event chemistry is of no help to him; he must depend on the sensitiveness of his eyes to colour.

White Mineral Pigments.—We are acquainted with a great number of white or, more properly, colourless mineral compounds; they possess the property of reflecting, undecomposed, all the rays of light which fall upon them, in consequence of which they produce that impression upon the eye which we call white. According as a white substance reflects every ray of light or absorbs a portion, we see it as a brilliant pure white or, in the latter case, as a white with a grey tinge. If a white body reflects the majority of the rays of light falling on it, but decomposes a small number, we perceive a white which has a yellow, blue or red tinge.

The most valuable white for the colour maker is evidently

that which reflects, unaltered, all the rays of light; it is the most brilliant, and free from every tinge of colour. The physical condition of the substance is most important. Solid substances are either crystalline, that is, possess definite shapes formed according to a regular law, or they are amorphous, that is, are composed of irregularly formed particles. Snow and white lead may serve as representatives of these two classes. Snow is composed of small colourless crystals of ice, the flat surfaces of which reflect, undecomposed, the light falling on them. The smaller are the crystals, the more pure appears to us the whiteness of the snow, and the thinner is the layer of snow required to produce the sensation of whiteness. But if the snow crystals are larger, the white appears to have a bluish tinge, and only a thick layer of snow is opaque. White lead, being an amorphous substance in a condition of very fine division, reflects the light very regularly, so that a thin layer of white lead appears quite opaque.

Among the artificial pigments, crystalline or amorphous, exactly the same conditions hold good as between snow and white lead. Of amorphous pigments a very thin layer is in most cases sufficient to make the surface upon which they are spread invisible, or, as the technical expression runs, "to cover," whilst crystalline substances possess a smaller covering power. A striking example of this is seen by a comparison of two white pigments, white lead and "patent white" (lead oxychloride). The former is amorphous, the latter crystalline. Both are completely colourless and reflect white light, but in consequence of its amorphous condition and finer particles, white lead possesses far greater covering power than "patent white".

Among all other colours the same rule holds. Amorphous pigments have always a greater covering power than crystalline. The smaller the crystals of the latter the greater

74   MINERAL AND LAKE PIGMENTS.

as their covering power, so that in preparing pigments of a crystalline character care must be taken to make the crystals as small as possible.

From the above definition of white pigments it follows that an immense number must exist, since every colourless substance in a state of fine division appears white. Generally only those bodies are used which are insoluble in water, or almost insoluble, and which possess great covering power. The following may be mentioned as white pigments, only a few of which are in use: white lead, white zinc, permanent white, lead oxychloride, lead sulphate and sulphite, zinc oxychloride, lead antimoniate, antimony white, tin white, tungsten white, and in addition certain earths, pipe clay, china clay, etc. Several of these pigments are far too expensive for ordinary use, and have no advantage over much cheaper pigments except for very special purposes, such, for example, as bismuth white for cosmetics.

In general use we find very few artificial white pigments; these are lead, zinc and barium compounds. Circumstances may arise which make it expedient for the colour maker to manufacture other white pigments, for example, a demand for them, or favourable opportunities for obtaining the requisite raw materials.

The white lead pigments, of which there is a large number, as we have indicated, all have the great disadvantage that they are not permanent, that is, are changed by atmospheric influences. It is well known that lead is a very delicate reagent for sulphuretted hydrogen, with the sulphur of which it forms a black compound. Now the air, especially of towns, contains sulphur in the form of sulphuretted hydrogen or ammonium sulphide; though the quantity is very small, the fate of every white or coloured lead pigment is decided by it; after a longer or shorter time it will be discoloured, will gradually darken, and finally be turned black. In spite of this great changeableness of lead pigments they are used by artists and painters, although the majority could be replaced by more permanent pigments entirely unaltered by the atmosphere.

WHITE LEAD.

This pigment, which in addition to other good properties has remarkable covering power, was amongst the earliest known artificial pigments. Already in the fourth century before Christ, Dioscorides described the preparation of white lead, which was obtained by exposing lead to the action of the vapours of vinegar, removing the white layer and treating it with water. The Roman writers describe a similar method; they use the name *cerussa*, under which white lead is known to-day in commerce.

Although white lead has been so long known, it was left to Bergmann, in 1774, to show that it contained carbonic acid; before that it was believed to be lead acetate. The development of analytical chemistry was followed by a knowledge of its constitution and the use of more rational methods of manufacture. Whilst in the middle ages the manufacture of white lead was almost exclusively in the hands of the Dutch and Venetians, in later years it gradually spread, and now many works are concerned in the manufacture of this pigment. That adulterations of white lead were not rare in former times appears from the writings of Basil Valentine, an alchemist of the fifteenth century.

Commercially white lead is known under most varied titles, of which the following are the principal: White lead, Venetian white, Dutch white, Krems white, Kremnitz white, flake white, etc.

According to its chemical composition, white lead is a compound of lead carbonate and lead hydroxide, that is, a basic lead carbonate. Commercial white lead, apart from intentional admixtures of other white substances, contains lead carbonate and lead hydroxide in varying proportions, as is shown by the following analyses by Mulder, who found that all the samples examined by him were composed according to one or other of the following formulæ:—

$2\ PbCO_3 . Pb(OH)_2$ containing 86·27 per cent. of PbO.
$5\ PbCO_3 . 2\ Pb(OH)_2$,, 85·86 ,, ,,
$3\ PbCO_3 . Pb(OH)_2$,, 85·45 ,, ,,
$4\ PbCO_3 . Pb(OH)_2$,, 85·00 ,, ,,

According to Hochstetter, the manufacture of white lead must be directed to obtaining the compound $2\ PbCO_3 .(OH)_2$, which possesses the following percentage composition:—

PbO . . . 86·32
CO_2 . . . 11·36
H_2O . . . 2·32

The compound of this composition is distinguished by being completely amorphous, and so possesses the greatest covering power. Commercial white leads, as is seen from the formulæ of Mulder, may differ appreciably from this composition, when they will have a smaller covering power, since they will contain some quantity of neutral lead carbonate, $PbCO_3$ (containing 83·46 per cent. of PbO), which is crystalline.

White lead is prepared according to very different methods, the principle of which consists in subjecting a solution of tribasic lead acetate to the action of carbonic acid, by which the basic carbonate is produced and the neutral acetate formed, the latter being then reconverted into the basic acetate, which again serves to produce white lead and so on.

CHAPTER VII.

THE MANUFACTURE OF WHITE LEAD.

THE processes by which white lead is or was manufactured may be divided, according to the principal operations, in the following manner:—

1. Processes in which metallic lead is subjected to the action of acetic acid vapour, whilst the vessel in which this operation is conducted is exposed to a higher temperature. In the oldest so-called Dutch method this increase of temperature is effected by the decomposition of manure, by which the vessels containing the lead and acetic acid are surrounded. In consequence of the heat produced by this fermentation, acetic acid and water are volatilised, and oxygen also being present, lead acetate is formed.

Also, as a consequence of the oxidation of the lead, heat is produced, which accelerates the process, and lead oxide is formed in large quantity, which unites with the neutral acetate already formed to produce a basic compound. The vessels in which this process is taking place are in an atmosphere containing much carbonic acid produced by the fermentation of the surrounding organic matter; this carbonic acid converts the basic lead acetate into white lead.

The German or Austrian method is to be regarded as an improvement on this rough process. The heat necessary for the normal course of the chemical reactions is produced from fuel; the carbonic acid formed by the combustion of the fuel is used to convert the basic lead acetate into white lead.

2. In the above methods the manufacture of white lead commences with the production of white lead from metallic lead and acetic acid. In the so-called French method a solution of basic lead acetate is decomposed by carbonic acid into white lead and neutral lead acetate, which latter is again converted into basic acetate.

3. The English method. The principle of this process consists in moistening litharge with a solution of lead acetate and exposing it to the action of carbonic acid, whereby white lead is formed.

Methods for manufacturing white lead, which are often advanced as entirely new processes, may be always traced to one of the above, from which they deviate but little in principle, and the deviations cannot always be regarded as improvements. In the following detailed account of the manufacture of white lead we shall adhere to the classification just given, according to which there are three principal methods: (1) Manufacture of white lead from metallic lead, acetic acid and carbonic acid (Dutch and German processes); (2) manufacture of white lead from basic lead acetate (French process); and (3) from litharge moistened with lead acetate solution (English process).

We should remark that every year "new" processes for the manufacture of white lead are patented. The majority of these will not be mentioned; to those acquainted with the principles of chemistry they at once appear impracticable.

Manufacture of White Lead from Metallic Lead.

(a) Dutch Process.

This essentially primitive process, when properly conducted, produces white lead of good colour and covering power, which are the properties for which this pigment is valued. It is now seldom used,[1] because other methods give a product of equal colour in a shorter time. It is, however, of economic interest, as showing how a branch of industry may rise from crude beginnings to a high state of perfection. The operations comprised in the Dutch process are as follows:—

(1) Casting the lead into sheets; (2) placing these sheets in pots and arranging the pots in the stacks (placing the pots containing lead and acetic acid in the bed of manure); (3) removing the pots from the stack; (4) separating the white lead formed in the pots; (5) further purification of the impure white lead by grinding, washing and drying.

1. Casting the Lead into Sheets.—At first sight it would appear unsuitable to cast lead in sheets, since this metal can be readily rolled into sheets of any thickness. Experience has, however, shown that rolled lead is only slowly attacked by acetic acid vapours, whilst cast sheets are rapidly attacked.

An iron pan about one metre in diameter, with an iron cover furnished with a pipe opening into a flue, is used for melting the lead. This arrangement is designed to protect the workman from the dangerous vapours evolved from the molten metal. At the workman's side of this cover is a counterpoised slide, which only remains open when the counterpoise is held in check by a lever. In front of the kettle there is an iron plate movable about a horizontal axis. The lead being heated to just above its melting point, the workman takes 7 to 8 kilogrammes of metal in a ladle, and pours it on the plate, which is horizontal. The lead solidifies in a very short time, but before it is completely solid the plate is inclined towards the pan so that the still

---

[1] In England the Dutch process is in general use.—TRANSLATOR.

80 MINERAL AND LAKE PIGMENTS.

liquid lead runs back into it, leaving a very thin sheet on the plate. The hard sheet is removed from the plate, and the latter cooled by cold water to be ready for a new casting. The sheets made in this way are not more than 1 to 2 millimetres thick. They are then cut into strips of a width to suit the size of the pots in which they are converted into white lead; the width of the sheets is generally 5 to 6 centimetres. Since the rate at which the white lead is formed depends on the surface of the metallic lead, instead of continuous sheets the lead is generally cast into gratings. For this purpose an iron plate, upon which are intersecting strips, is used instead of the flat plate. Plates for the cast-

Fig. 4.

ing are also used containing grooves intersecting at right angles. In the first case, plates are obtained in which are openings meeting at right angles, and in the second case, according to the distance of the grooves apart, a more or less wide-meshed lattice work.

2. **Building up the Stacks.**—The rolled-up lead plates are placed in the pots. These (Fig. 4) are somewhat conical in shape; they have at some distance from the bottom a projecting ring, or sometimes three projections only, upon which the lead spiral rests. Before the spirals are put in position, a quantity of ordinary vinegar, about a quarter of a litre, is poured in. There must be sufficient room below the spiral so that it shall not be in contact with the vinegar.

The insides of the pots are glazed at least half-way up, so that the liquid does not penetrate the porous earthenware.

The pots have a capacity of about 1 litre, and a diameter at the top of 10 centimetres. If lead plates are used, the pots are about 20 centimetres high; if gratings are used the pots may be lower, by which there is economy in room and a larger number can be placed in one stack.

The stacks, built up of pots and manure, are of different sizes; it is not advisable to make them too small, or the loss of heat would be considerable. A stack 4 to 5 metres long, $3\frac{1}{2}$ metres wide and 6 to 7 metres high will contain 6,000 to 8,000 pots and 9,000 to 11,000 kilogrammes of lead.

The stack consists of a rectangular pit walled on three sides; the fourth side is open, with the earth dug out in the form of an inclined plane, in order to permit the introduction of the pots and the manure. The construction of the stack is commenced by placing the pots at the bottom in rows, avoiding interspaces as much as possible. In a stack of the size mentioned a layer contains 1,000 to 1,200 pots. Between the pots containing lead and acetic acid are arranged a number of larger ones containing acid only; the object of these is to furnish acetic acid vapour. When the pots are in place, 3 or 4 lead plates are placed on each spiral, the top plate forming the cover; immediately over the pots strong wooden planks are laid, and upon them a layer of boards, which must fit so tightly that nothing can fall through. On the boards is spread out a layer of fresh stable manure, with which the space between the outside row of pots and the wall is also filled. The layer of manure is 30 to 40 centimetres thick.

Upon the lowest layer of pots, a second, third and so on are built up exactly in the same manner, so that the whole stack is filled with alternating layers of pots and manure. In order to prevent the cooling of the uppermost layer of pots, it is covered with a thicker layer of manure 60 to 70 centimetres thick. When lead plates and the taller pots are used, a stack will generally contain 15 layers; but when gratings and the smaller pots are employed, 18 layers can be packed into the same space. In arranging the layers of pots, care should be taken to leave spaces at tolerably equal distances, so that the air necessary for the oxidation of the lead may enter. To prevent the cooling of the stack at the front, where it is not protected by masonry, when full it is walled up with boards; a board roof also protects the erection from rain.

In place of manure, spent tanners' bark can be used, which in the same manner ferments, producing heat and carbonic acid. In places in the neighbourhood of large tanneries this spent bark is generally obtainable at lower prices than stable manure, which is more valuable for agricultural purposes; the former has also a very considerable advantage, white lead made by means of spent tanners' bark being generally of a purer white than that made with manure. The reason for this is that in the decomposition of animal excrement small quantities of sulphuretted hydrogen are produced, a gas which produces black lead sulphide when it comes into contact with lead compounds.

According to the results of practical experience, pigs' dung cannot be used in the manufacture of white lead; so much sulphuretted hydrogen is evolved from it that the white lead is not white, but has a greyish tinge.

When bark is used in place of manure, a discolouration of white lead by sulphuretted hydrogen is not to be feared, but there is the drawback that a longer time is necessary for the corrosion of the lead, because the bark decomposes more slowly than the manure, and accordingly gives out less heat and carbonic acid.

According to the climate of the country, the stacks may

be differently erected. In colder countries it is necessary to sink them in the earth and surround them with masonry, as directed above; but in warmer climates such effectual protection against cooling is unnecessary, yet in all cases it is better to sink the stack in the earth on account of the regularity of temperature so obtained.

Instead of sinking the stacks in the earth, they may be built in the open when there is a plentiful supply of manure or bark; but they must then be surrounded by a very thick layer of manure to prevent cooling. It is a desirable alteration in the construction of the stacks to provide the pots with lids, and so avoid the use of the layer of boards separating each two layers of pots. The object of the cover is simply to prevent dirt from falling into the pot. It should not fit tightly on the edge, or the entry of carbonic acid into the interior of the pot would be made difficult. The lids are therefore rounded and fit loosely on the pot. When pots with lids are used, the lowest layer is covered with manure in the ordinary way; upon this again comes a layer of pots, and so on.

The transformation of the lead may be regarded as complete in four to six weeks when manure is used, but with bark the time extends to ten weeks. The quantity of white lead obtained varies in different cases; for example, from a stack 5 metres long, 4 metres wide, and 6 metres high, into which 12,000 kilogrammes of lead were put, 10,000 kilogrammes of white lead were obtained, and 4,000 kilogrammes of lead remained unaltered. In another case, for a stack of 8 layers 280 kilogrammes of vinegar and 9,600 to 12,000 kilogrammes of lead were used, and there was a residue of 10 to 15 per cent. of lead.

3. **Removal and Grinding of the White Lead.**—The stack is pulled down after the lapse of the necessary time; the lead plates and rolls are collected in wooden boxes and brought

into the room where the white lead is separated from the metallic lead. Formerly the white lead was removed from the sheets exclusively by manual labour, an operation extremely dangerous to the workman. It is quite impossible to prevent the formation of white lead dust, so that the men were continuously in an atmosphere charged with the poisonous material, and, as a consequence, suffered from the various forms of lead poisoning.

In order to diminish as much as possible the injurious effects of white lead on the health of the workmen, manual labour has been, as far as possible, replaced by machinery, yet the greatest care is necessary in the different manipulations of so poisonous a substance as white lead.

When the white lead is removed by manual labour, an operation which ought to be forbidden, the lead sheets are unrolled and struck together, whereupon the greater part of the white lead falls off. To remove the remainder of the white lead the plates are laid one upon the other and struck with a hammer until the white lead is loosened; or the plates are cleaned with metal brushes.

The masses of white lead obtained in this way are contaminated by larger or smaller quantities of metallic lead, from which they must be freed by a further mechanical operation, namely grinding. The larger pieces of the white lead, which have a thickness of several millimetres, were picked out and sold separately under the name of flake white. This was formerly a highly prized quality of white lead, its appearance being a guarantee of its purity. The flake white generally found in the market nowadays is not obtained in this manner, but by mixing white lead with a solution of dextrin, forming plates from the paste and drying them slowly in the air.

In order to separate the white lead mechanically from the remains of the lead sheets, grooved rollers are used, between which the sheets are passed. In order to guard against dust, the rollers are surrounded by a closed casing, in which there is also a sieve which serves to separate the larger pieces of white lead (flake white) from the fine dust. The arrangement is represented in Fig. 5. The unrolled plates pass through the opening, B, on to an endless leather band, by means of which they are carried between the grooved rollers, D and E; after they have passed through these they go beween a second pair of rollers, F and G, which are placed nearer together; they then fall into the

Fig. 5.

drum-shaped sieve, H, out of which they leave the apparatus in the direction of K. The white lead falls through the sieve, is caught in trucks placed at J, and carried away to the mills.

A mechanical arrangement for the separation of white lead from the unaltered metallic lead, due to Horn, is represented in Figs. 6 and 7. It consists of a drum, in which is the spindle, b, provided with the arms, c. The teeth, e, on the arms reach nearly to the lower curve of the drum, but pass at a somewhat greater distance from the upper portion so that the pieces of lead can fall down again after they are carried up. The material is fed in through the hopper, *t*, the rotation of the arms loosens the white lead from the pieces of metallic lead, and it is carried on by the water which passes through the sieve, 3, and the stop cock, *h*, into the settling tank. The lead pieces which are carried forward have their progress checked by the bridge, *o*, they are collected by the perforated scoop, *n*, are raised above the hopper, *m*, and fall out of the apparatus at *r*. If it is required to continue the treatment of the material in the drum for a longer

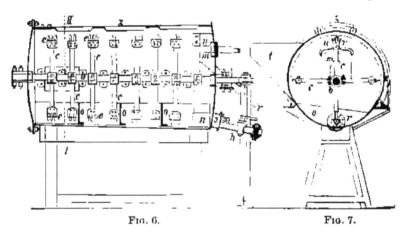

Fig. 6.   Fig. 7.

period it is only necessary to close the hopper, *m*, by the slide, *u*.

White Lead Mills.—Before the white lead is subjected to the real process of grinding, it is generally first ground dry, or, more properly, pressed or crushed. This crushing is accomplished by means of vertical or edge-runner mills, which consist of mill-stones running round upon a stone bed about a vertical axis. The mill is surrounded by a wooden casing to prevent the escape of dust.

Wet grinding, which is done between mill-stones, may be carried out in two ways, with the production of hard or soft white lead. The former is obtained when the lead acetate is not removed, the latter when all the lead acetate is washed out.

Hard White Lead presents a shining mass, broken with difficulty. This appearance is a guarantee against adulteration with barytes; white lead, which contains this adulterant,

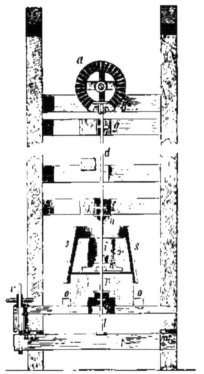

Fig. 8.

does not give a smooth, but an uneven earthy fracture. Hard white lead is rather difficult to grind, and requires very careful treatment to be brought into that state of fine division in which it is usable.

The dry white lead, after powdering under the edge-runners, must pass through a sieve, which retains the par-

88 MINERAL AND LAKE PIGMENTS.

ticles of metallic lead, before it is subjected to wet grinding. The mills used for wet grinding differ little, or not at all, from the ordinary pattern.

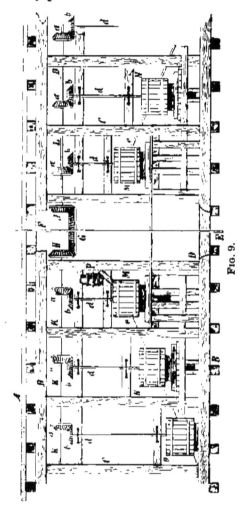

Fig. 9.

A mill designed by Richter of Königsee, in Thuringia, for wet grinding, is constructed as follows (Figs. 8 and 9). The shaft, FE, is driven by a water-wheel and the necessary

THE MANUFACTURE OF WHITE LEAD. 89

gearing at 60 revolutions per minute; the bevelled cogwheel G, by means of the cogwheels H and J, drives the shafts K and L, which, by means of the bevelled cogwheels a and b, communicate to the runner stones of the mills, M, N and O, a speed of 140 revolutions per minute. C D is the framework for the support of the storied arrangement of mills, f and g are bearings with cast-iron cups to receive the escaping oil, i is a clamp fastening the axle d to k. A steel cup in k rests upon the hardened top of the spindle, l; the bed stone rests upon the beam n (80 millimetres thick) and is prevented from lateral movement by the surrounding oo; p is a wooden support in the opening in the bed stone, carrying the stuffing box, q, which prevents the white lead from running out, and connects the spindle, l, with the bed stone. The runner stone, r, is surrounded by the box, s, cemented with white clay inside and out and fastened to the bed stone. The lever, t, turning about w, moved by the screw, u, and handle, v, regulates the position of the runner. The diameter of the stones is 95 centimetres.

Thirty kilogrammes of white lead, mixed with water, are fed into the top mill of the series shown on the right; the mixture flows by the spouts, e, into the lower mills, and is received in vessels, from which it is emptied into the holder, P. From P it goes through the 3 mills on the left. After passing through 6 mills, white lead. without barytes is ready for use. If the white lead contains an admixture of barytes, it is put through the 3 mills on the left a second time. In 24 hours 900 to 1,200 kilogrammes (18 to 24 cwt.) of white lead are ground.

In dry grinding, mills are used arranged so that the formation of dust is avoided, a matter which is of particular importance. The usual size of these mills has a diameter of 90 to 95 centimetres for both runner and bed stone. The grinding surfaces of both stones have radial grooves.

Fig. 10 represents the construction of Lefèbre's white-lead mill, which is designed to give the greatest protection possible to the workmen. At A the white lead masses are enclosed in a hopper lined with bronze, with internal angular projections; by means of M, which is grooved in a similar manner, the larger lumps are broken up and enter the mill

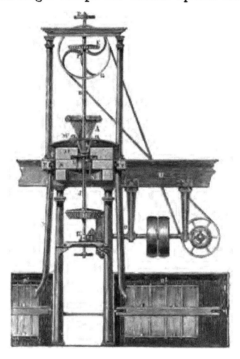

Fig. 10.

itself by means of the hopper under A. The mill consists of bed stone, H, and runner, K. The grinding surfaces are grooved to facilitate the delivery of the ground white lead. As the illustration shows, the stones are completely covered by M, so that the escape of dust is almost completely prevented. The ground white lead is conducted by the tubes, O O, to the drums, O' O', from which it is packed.

In order to obtain hard white lead, the material which has been ground dry under the edge runners is mixed with water to a soft paste, or if water has been added whilst the lead was under the edge stones, more is now added to make the paste thin enough to be ready for the mills, where it is ground, being fed in regularly by means of a copper spoon. The paste issuing from the mill is collected in earthenware pans or plaster of Paris moulds, in which it is dried.

These pans have usually the shape of a truncated cone, which was the form in which white lead was formerly brought into trade from Holland. In drying, water is lost, and the mass of white lead shrinks, so that the lump can be removed from the mould after a few days on turning it upside down. The drying is accomplished either in the air or in artificially heated stoves. The heating must at first be gradual, or the mass of white lead would shrink so rapidly that the cone would be full of cracks, and then easily fall to pieces. When once the drying has reached a certain stage, the temperature of the stove may be raised to 50° C. (122° F.) without danger of breaking the lumps. When quite dry, the surface of the white lead, which is now rough, must be smoothed by scraping, when it is ready for the market.

Soft White Lead.—The hard white lead prepared as just described consists of very heavy and very hard lumps of the purest white. When soft white lead is required, the admixed lead acetate must be removed by washing. This is accomplished by adding a larger quantity of water, either when grinding under the edge runners or in the mills, so that a thin pulp is formed; this is run into a receiver, in which is a stirrer. The white lead is not completely prevented from sinking by the stirrer; a soft mud is deposited at the bottom of the vessel, which can only be stirred up with difficulty on account of its high density. When this vessel is full, the stirrer is stopped and the milky liquid allowed to settle, which happens in a short time on account of the high specific gravity of white lead. The clear liquid above the white lead is drawn off into a tank lined with cement. It is advisable to arrange the stirrer so that it may be placed at any height in the vessel; if this is the case, by gradually lowering the stirrer whilst in motion, the white lead lying at the bottom can be mixed up with fresh water. These operations are repeated until all the lead acetate is removed.

Soda solution is added to the wash waters in order to recover the lead dissolved in them as lead acetate; lead carbonate is precipitated and settles at the bottom of the tank. This precipitation may also be effected by putting lumps of limestone in the tank. The paste remaining in the washing tubs, which now contains only pure white lead and water, is filled into bags of closely woven material and the water pressed out by a gradually increasing pressure, until a stiff pulp remains behind. This is then completely dried, either in the air or in drying stoves.

Soft white lead forms either irregular lumps or a soft, heavy powder. The lower qualities of white lead contain a smaller or larger quantity of finely ground barytes; the higher the proportion of barytes the smaller is the covering power of the mixture. A very simple method for detecting barytes in white lead will be given later.

(b) German Process.

The German or Austrian method of making white lead is also known as the chamber process, since the formation takes place in closed chambers, constructed of wood or masonry. In the older processes, cast lead sheets were bent double and hung on cross bars in a wooden box with a watertight bottom, with the precaution that the plates did not touch. A number of these boxes, generally 90, were arranged in a hot room, each box being about 1·6 metre long, 0·4 metre wide and 0·3 metre high. On the bottom of each box was poured a mixture of vinegar or dilute acetic acid and wine refuse, and the box was covered with a well-fitting lid. The temperature of the room was gradually raised week by week; during the first week remaining at 25° C., during the second at 38° C., the third at 45° C. At the commencement of the fourth week the temperature was raised to 50° C., at which it was kept for a fortnight. At this high temperature a considerable quantity of acetic acid is evaporated, causing the formation of lead acetate, which is converted, by the carbonic acid evolved from the wine refuse, into basic lead carbonate. When the proper temperatures have been maintained, on opening the boxes almost all the lead is found changed into white lead, which is knocked off, and the residual lead used in casting new plates. It is easy to conceive that the boxes may be well replaced by brick chambers, in which a large number of lead plates are brought and into which acetic acid vapours and carbonic acid are introduced after the room has been closed. Chambers are used which can contain 12,000 to 12,500 kilogrammes of lead.

These chambers, in which the lead plates are hung upon wooden supports, have an opening immediately above the bottom, which is connected with a retort in which vinegar, containing 4 to 5 per cent. of acetic acid, is boiled. After about 12 hours, by the simultaneous action of the vapours of acetic acid and the oxygen of the air, lead acetate is formed. Carbonic acid is now led into the chambers. The carbonic acid is obtained by burning charcoal in a cylindrical furnace; it is cooled by being passed through a long iron tube before it enters the chamber. For 12,500 kilogrammes of lead there are required every day about 482 litres of dilute acetic acid, obtained by mixing strong acetic acid with water until the mixture contains 4·5 per cent. of the acid, and 18 kilogrammes of charcoal, from which the carbonic acid is obtained. The time required is 5 or 6 weeks, and the residue of unaltered lead varies from 10 to 35 per cent.

The white lead obtained by this process is quite usable, but the process has the considerable disadvantage that there is no control over the quantities of materials used. In order to produce a certain quantity of white lead of a certain composition, definite quantities of lead, oxygen, acetic acid and carbonic acid are necessary. If, then, the apparatus can be so arranged that the quantities of material employed can be accurately measured, a great advance will have been made, for the operation will be no longer conducted at random, but under definite unchangeable conditions. The quantities of carbonic acid and acetic acid can be calculated beforehand. The volumes of the gases required may be measured without difficulty by meters of the type used for measuring coal gas. Upon this principle are based a number of more modern methods.

According to the process of Major, the vapours of water and acetic acid are introduced at the same time into the chambers filled with lead plates, in order to produce basic lead acetate. This part of the process requires about 12 hours. Then carbonic acid is introduced; it is produced by burning charcoal in an iron cylinder, through which air is forced. By means of this carbonic acid, which is at a comparatively high temperature, about 60° C., the lead acetate is quickly changed into white lead. A portion of the lead acetate remains undecomposed; in order to remove this, at the close of the operation ammonia is injected, which decomposes the lead salt. The ammonia salts now present are finally driven out by means of superheated steam. Major's apparatus is depicted in Fig. 11. A and B are chambers with horizontal gratings upon which the lead lies ; C is the furnace for burning the charcoal, provided with a fan. The products of combustion from C pass under the boiler D, their passage being regulated by the valve, a; they then heat the boiler E,

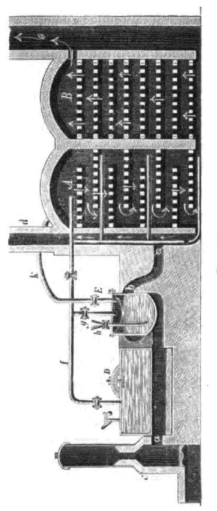

Fig. 11.

containing acetic acid, and enter the chambers, A and B, through the flue, b, or are directed by the valve in b into the chimney, d. The gases, after passing through the chamber, find an exit at e. The steam pipe, f, conducts steam into the boiler E, through g, and also into the chambers, A and B, thus heating them and providing the necessary moisture; h is the funnel for filling E; the pipe, k, carries the acetic acid vapours into the chambers. After the thin lead plates have been brought into the chambers they are closed and the temperature raised to 49° to 60° C., steam and acetic acid vapours are led in for 10 to 20 hours in order to form the basic acetate, then the furnace gases are introduced at a temperature of 60° C. The white lead obtained in this manner can be finished in the ordinary way by washing and grinding, but it is better to

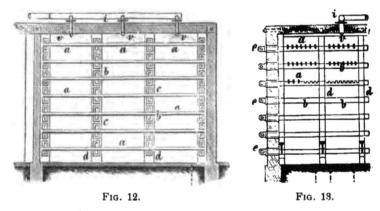

Fig. 12.   Fig. 13.

remove the lead acetate by introducing ammonia, and then hot air or superheated steam, as previously stated.

The process is complete in 2 to 4 weeks. Gartner, working according to this process with 150 kilogrammes of lead in a chamber 1·26 metre long, 0·78 metre high and 0·78 metre wide, obtained good white lead in 28 days.

The apparatus designed by H. Kirberg for the manufacture of white lead is illustrated in Figs. 12, 13 and 14. The lead plates are hung upon the laths, a, in the chambers; the supports of the laths, b, go through the slits, c, in the supports, d, and project through the walls at e. The carriers, b, hang from bolts by brass wire. By striking the end of the laths, e, they are made to swing so that the white lead loosely adhering to the lead plates is shaken off and fresh surfaces of metal are exposed. The openings in the walls through which the ends of the laths project are closed by indiarubber. In order to prevent the production of dust when the chambers are emptied, water is introduced in a fine spray upon the lead plates from the copper tubes, v, placed under the roof of the chamber. The water washes off the remains of the white lead from the plates.

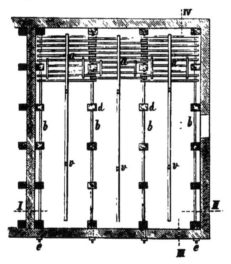

Fig. 14.

In a similar manner, the white lead process may be carried out in a shorter time when the gases enter the chamber under increased pressure; but this is attended with difficulties, since continuous supervision of the apparatus and of the tightness of the chamber is necessary.

The apparatus of W. Thompson (Fig. 15) consists of a chamber, A, constructed with a false roof, B, to carry off the condensed vapours, and provided at both sides with doors, through which the waggons, carrying the lead, are introduced and removed. The pipes, n, which convey air and carbonic acid, are provided with branches, e, which reach to the sides of the chamber, and are joined to the perforated tubes, D. The pipes, n, are provided with reservoirs, b, which prevent the too rapid entry of the air or carbonic acid, and at the same time give a sufficient heating surface. Upon the waggon, E, are the frames, v, which support the lead plates. The plates may be solid or gratings; they are 3 to 12

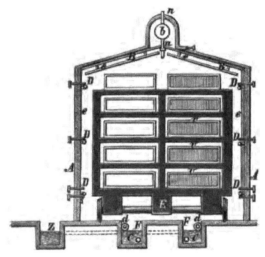

FIG. 15.

millimetres thick, and are arranged at intervals of about 25 millimetres in the frames. The troughs, F, for the reception of the acetic acid are filled from the reservoir, Z. The steam pipe, C, effects the evaporation of the acid; the steam pipe, d, heats the chamber at the commencement of the process, and also when C is out of use. When the chamber is filled with lead, the troughs, F, are filled with acetic acid of 5 per cent. strength. The chambers are closed, and steam is led through C and d until the temperature in the chamber reaches 25° to 50° C., and the evaporation of the acid begins. Then air is forced in under a small pressure through n, e, D during 3 to 4 days, after which follows a mixture of equal parts of carbonic acid and air, continued until the formation of white lead is complete. When plates 3 millimetres thick are used, and the temperature is gradually raised from 25° to 50° C., about 12 days are required; but if the plates are 12 millimetres thick, 28 days are necessary. For the proper carrying out of this process it is important that the temperature of the chamber should be very gradually raised from 25° to 50° C.

In the process of P. Rey molten lead is poured in a thin stream into water, and the "granulated" metal then placed in vessels, in a layer 30 centimetres thick, upon a grating 5 centimetres above the bottom. In the bottom are narrow tubes which, reaching above the lead, serve to admit air. Acetic acid is allowed to flow over the lead from vessel to vessel. In order to obtain the proper solution, a layer of lead, 2 metres thick, is necessary, so that 6 to 7 vessels are arranged, one above the other. If lead acetate solution be used, the layer of lead need only be 1·2 metre thick. The solution of basic lead acetate is then treated with carbonic acid according to the ordinary process.

In addition to the modifications of the German white lead process which have been described, many others have been proposed. The principle of many of these methods is, that finely divided lead is much more rapidly converted into white lead than lead in the form of plates; finely divided lead exposes an enormously greater surface to the action of the acetic acid and other materials than do lead plates.

In Rostaing's method the lead is changed into very small pellets, by allowing the melted metal to flow on to a rapidly rotating iron disc; the molten lead, in consequence of the centrifugal force due to the rapid rotation, assumes the form of very small drops, which are thrown off the disc and cooled in a vessel of cold water.

Torassa recommends that lead, obtained in pellets by pouring into cold water, should be brought into a rapidly rotating vessel, whereby the greater part will be converted into fine dust (?). This lead dust is said to be then converted into white lead by simple exposure to air, lead oxide being first formed, and from it basic lead carbonate. This process is not workable on a large scale. The rapid oxidation of finely divided lead by air has been applied by several inventors to the manufacture of white lead. The processes of Woods, M'Cannel and Grüneberg are founded upon this transformation. In essentials they are as follows: lead is finely divided in iron or earthenware cylinders, and either carbonic acid and air, or a mixture of these with acetic acid vapour, are introduced by means of the axle of the cylinders.

If a special apparatus is not provided, as it should be, on the large scale, to separate the unaltered lead from the white lead, the lumps coming from the chambers must be subjected to a process of levigation, in which the lead, being the heaviest body, will be first deposited; the last portions deposited consist of the purest white lead, the lower layers of which will be tinged more or less grey by an admixture of finely divided lead and lead peroxide.

White lead manufactured by the German process occasionally exhibits perceptible reddish or greyish tinges, the cause of which lies in the defective execution of the process. The red tinge denotes the presence of free lead oxide, caused by the use of an insufficient quantity of acetic acid. A grey tint is due to the presence of metallic lead or an excess of lead carbonate.

(c) French Process.

The French process for making white lead is based upon the reaction which occurs when carbonic acid is passed into a solution of basic lead acetate; basic lead carbonate is precipitated and neutral lead acetate remains dissolved; the latter is again converted into basic acetate, from which carbonic acid again separates white lead and so on. This process, at present used on an enormous scale, is due to the French chemist Thénard, who first put it into operation on the large scale at Clichy, near Paris. The method is also known as the Clichy process.

The operations of this process are divided into the production of the basic acetate and the treatment of its solution with pure carbonic acid, whereby the basic carbonate is precipitated.

1. Preparation of the Solution of Basic Lead Acetate.—The preparation of this compound has been already described; the following is supplementary to what was previously given. If litharge be used, it is dissolved in wooden tubs heated by steam. The acetic acid is brought nearly to boiling by open steam, and the finely ground litharge gradually added. In consequence of its high specific gravity, the litharge would quickly sink to the bottom, so that it is advisable to keep the liquid in motion by means of a stirrer, and to allow the litharge to fall in in a thin stream. The introduction of the latter is continued until the specific gravity of the solution indicates that the liquid contains three equivalents of lead oxide to one equivalent of acetic acid.

In working with metallic lead, this must be used in a finely divided form; it is cast, as in the Dutch process, into thin sheets or gratings, or into flat wires or ribbons. These ribbons are easily made by melting the lead in a pan provided with a delivery pipe with stop cock; beneath the latter is brought a vessel, which can be moved backwards and forwards upon a tram line, filled with water. When the melted lead is allowed to flow into this vessel, which is being moved backwards and forwards, the metal forms long, thin ribbons, which possess a large surface. A wooden tub is almost completely filled with these lead ribbons, which are then covered with acetic acid. After a short time the acid is run off, when, by the action of the air, so energetic an oxidation of the lead takes place that the contents of the tub become heated, and steam and acetic acid vapours begin to rise. When this is seen, the original acetic acid is pumped back into the tub, and left there for some hours in contact with the lead in order that it may dissolve the lead oxide. When the solution has reached a specific gravity of 1·1326 to 1·1415, it is drawn off from the undissolved lead, which again, in a short time, in consequence of the rapid oxidation, becomes warm, and is treated with fresh acetic acid.

The lead ribbons become finally so thin that they fall together by their own weight, and form tight masses, upon which air and acetic acid can no longer act. These residues are then removed from the solution vessel and new ribbons introduced. The residues have a velvety appearance; when they are mixed up with water they make it dark, and from the turbid liquid a fine, velvet-black powder soon separates, which consists of finely divided silver. The liquid still remains turbid owing to the suspension of fine particles of carbon. The lead ores, especially galena, often contain notable quantities of silver; silver lead is generally desilverised before use; usually, however, small quantities of silver remain in the lead. When the lead is dissolved in acetic acid, this silver settles as a soft powder at the bottom of the vessels in which the lead residues from the solution tubs are washed.

2. **Preparation of the Carbonic Acid and Precipitation of**

the White Lead.—The carbonic acid required for precipitating the basic carbonate is obtained either by heating limestone in a small furnace, from which the gas is drawn by a pump, or directly by burning charcoal. In the former case very pure carbonic acid is obtained, and, as a by-product, valuable quicklime; in the latter case precautions must be taken to produce pure carbonic acid.

The furnace designed by Kindler for preparing carbonic acid (Fig. 16) consists of a conical furnace, burning coal or coke upon the hearth, a. The passage, c, is divided by a vertical wall, in order to avoid obstruction from the piling

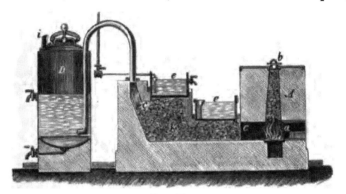

Fig. 16.

up of fuel. The space, K, is filled with limestone, through which the carbonic acid passes; the tanks, e, filled with water, cool the limestone, and the gas, which then passes through the water in the washing vessel, D, is drawn off by a pump.

In the old process at Clichy the apparatus depicted in Fig. 17 was used. The basic lead acetate was made in the wooden tub, A, provided with the stirrer, B C. The solution was run off from this vessel by means of the cock, b, into the settling tank, E, in which the mechanical impurities separated from the solution. The clear liquid ran into the decomposing vessel, a large shallow covered tank, holding 9,000 to 10,000 litres. In this tank opened 800 copper tubes, given off from the large pipe, S. The small furnace, D, in which limestone was burnt with coke, produced the carbonic acid; from the pipe at the top of the furnace the carbonic acid was brought into the Archimedean screw, h K, was washed with water, and pumped into the solution of lead acetate. The introduction of the carbonic acid was continued from 10 to 12 hours, after which the apparatus was left at rest until the liquid in the decomposing tank had become quite clear, through deposition of the white lead. The clear solution, now containing neutral lead acetate, was

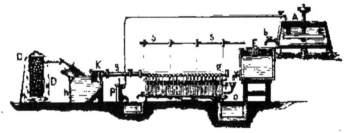

Fig. 17.

run off into the receiver, m, from which the pump, P, carried it back into the dissolving tub, A, where it was treated with fresh quantities of litharge. The solution of neutral lead acetate drawn off from the white lead had approximately a specific gravity of 1·0901. The white lead at the bottom of the decomposing tank was a tolerably thick paste. It was transferred to the tank, O, and washed several times with water. The first wash waters, which contained small quantities of lead acetate, were returned to the dissolving tub. The resulting white lead formed a very soft powder; it was at once placed in the drying pans. The white lead prepared by this process is a precipitate containing no coarse lumps, so that grinding is unnecessary.

Theoretically, the quantity of acetic acid with which the process is commenced is sufficient to form an unlimited quantity of white lead, since all the acetic acid brought into the decomposing tank in the form of basic acetate is returned to the dissolving tub as neutral acetate. In practice, however, matters are somewhat different. Small quantities of acetic acid are lost in the wash waters; each time the solution of lead acetate is pumped back into the dissolving tub, a small quantity of acetic acid must be added to make up the loss.

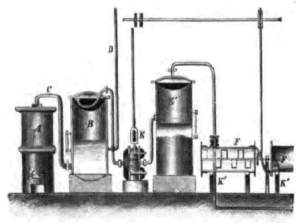

Fig. 18.

The method pursued by Ozouf, in France, is a considerable improvement on Thénard's process. Pure carbonic acid is used for the precipitation, and white lead of similar composition to that produced by the Dutch process is obtained, since the introduction of the carbonic acid can be regulated according to the volume and strength of the lead solution, and thus white lead of any desired composition can be produced. The most elaborate precautions for the health of the workpeople are taken.

The preparation of pure carbonic acid gas is based upon

the absorption of this gas from a mixture of gases by a solution of sodium carbonate, and its evolution on heating the solution. The apparatus is shown in Figs. 18 and 19. The products of combustion obtained from the stove, A, are drawn by the air pump, E, through the pipe, C, into the cooler, B, which is regularly fed with cold water by D. The gases compressed in the receiver, E', deposit moisture there, and then proceed through 3 horizontal cylinders, F, of sheet iron, provided with agitators, in which the carbonic acid is absorbed by a cold solution of sodium carbonate of 9° B. The unabsorbed gases escape into the atmosphere through G (Fig. 19). The sodium bicarbonate solution is received in the wooden tank, H, after passing through the 3 cylinders, F. The pump, I, of the alternating pump, $I\ I'$, lifts the sodium bicarbonate solution out of H and sends it through the pipe, K, into the tubular cylinder, J, which stands upon a cylinder of larger diameter, J', communicating with it only by the vertical tubes. The bicarbonate solution rises between the tubes in J, passes through the pipe, L, drops in a fine spray through the rose forming its mouth, and by means of the vertical tubes passes into J' and thence into M, where it is heated by means of a steam coil to 100° C. Carbonic acid is then evolved, and the residual solution of neutral sodium carbonate, after cooling in the vessel, R, by means of the cold coil, is drawn off by the pump, I', again to enter the cylinders, F, by means of the pipe, K. The carbonic acid evolved in M, together with steam, enters J' through N, and in rising in the tubes of the cylinder, J, is cooled by the falling spray of bicarbonate solution. The cooling is completed in the coil, O, surrounded by water; the vessel, P, separates the condensed water and passes the gas on into the holder, Q. The pipe, S, connecting P with the suction pipe of the pump, J', serves to restore to the solution of sodium carbonate the water it has lost, thus maintaining the proper concentration. The cost of 1 cubic

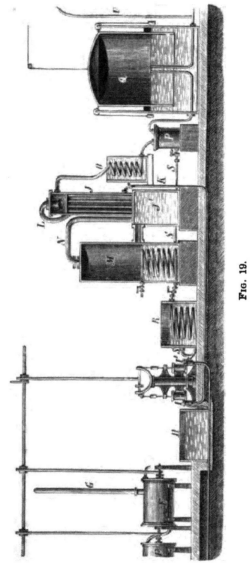

Fig. 19.

metre of carbonic acid is 10 centimes, of 1 kilogramme 5 centimes.

For the production of white lead, the carbonic acid, by means of the pipe, U (Figs. 19 and 20), enters the cylinder, T, provided with an agitator and containing a solution of basic lead acetate. By means of the pump, V, the lead solution is fed into the cylinder, T, through W. The absorption of the gas proceeds rapidly; the progress of the operation is followed by the observation of a pointer moving over a scale; as the gas holder sinks the pointer moves upwards. After the precipitation of white lead, the contents of T are emptied into the tub, b, in which rotate rakes attached to a vertical axis of iron plated with copper. When the white lead has settled, the supernatant solution of neutral lead acetate is drawn off through the pipe, c, by means of the pump, d, and conducted into the water-tight vessel, X, containing a stirrer on the vertical axis, W, made of coppered steel. Here litharge is added, and the resulting solution of basic lead acetate is conveyed to the cylinders, T, by means of the pump, V, as already described. The white lead in the tub, b, by putting the stirrer in motion, is washed once with water which has been previously purified by a little lead acetate. It then goes into another tub provided with stirrers, where it is several times washed, sodium carbonate being added to the last wash water until a sample of the white lead is not coloured by a drop of potassium iodide solution. In this way the wash water is obtained free from lead, and the product is said to be of better quality. This, however, is not in accordance with the fact that good Dutch white lead generally contains some lead acetate. The two-cylinder pump, h, which is in connection with the gas holder, forces gas over the surface of the liquid in T in order to drive it into tubs which are not in the position shown for b in the illustration, and into which there is no direct flow. The washed white lead is brought into bags which are pressed in a hydraulic press, dried, ground, sieved and packed in casks.

THE MANUFACTURE OF WHITE LEAD. 109

These troublesome and often dangerous operations have been

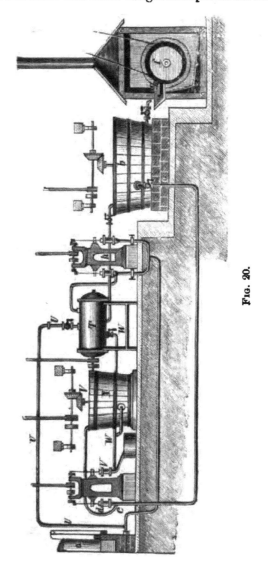

Fig. 20.

modified by Ozouf in the following manner. The pulp white lead runs from the tub, *b*, into the hopper, *g*, where it is kept mixed by a small stirrer, and from which it passes on to the cylinder, f, heated by gas from the inside. In its rotation the cylinder carries along the white lead and dries it, it is then removed by a knife below the hopper, and falls on to an inclined plane. The hopper and cylinder are in a room provided with a good draught.

The lumps coming from the drying room are placed by workmen wearing respirators in buckets on an endless chain, are carried to the mills, ground and sieved; then, by means of an Archimedean screw, the white lead is conveyed to a cask in which it is evenly pressed by means of a special mechanism. A bell announces when a cask is full.

Manufacture of White Lead by means of Natural Carbonic Acid.—In districts where currents of carbonic acid gas issue from the ground, they can be used in the manufacture of white lead, and are actually utilised for this purpose. Natural carbonic acid may, of course, be used for any of the white lead processes.

(d) ENGLISH PROCESS.

In this process, now no longer in use, white lead was obtained by mixing litharge to a stiff paste with a weak solution of lead acetate and exposing the paste to the action of carbonic acid. By continually kneading the mass by means of grooved rollers or of rotating cylinders, through the hollow axis of which carbonic acid was led, the paste was thoroughly brought into contact with the carbonic acid.

By this process a good product is only obtained when pure litharge, entirely free from the oxides of iron and copper, is used. The copper oxide may be removed from the litharge by means of ammonia if this can be obtained at a low price; but oxide of iron cannot be removed, and very small quantities of it are sufficient to impart a yellow tinge to the white lead.

(e) Other Methods.

In Payen's process the lead sulphate obtained in considerable quantities as a by-product in calico printing is the raw material employed. By treating this lead sulphate with a solution of ammonium or sodium carbonate, white lead and ammonium or sodium sulphate are produced. The white lead is then freed from the soluble salts by washing, mixed with a small quantity of lead acetate, and pressed into the drying moulds.

By boiling lead sulphate with caustic soda and passing in carbonic acid (Puissant's process), a white lead is obtained which differs considerably in composition from ordinary white lead.

Many methods have been proposed with the object of converting insoluble lead salts, obtained as by-products or by an inexpensive process, by treatment with alkaline or alkaline earth carbonates, into white lead. The fact that none of these methods has obtained a permanent footing in the industry shows that each must be accompanied by serious defects, or can only be practicable under peculiar conditions.

Magnesium carbonate is used in Pattison's process to decompose lead chloride. Dolomite (magnesian limestone) is the raw material for the magnesium carbonate. Coarsely powdered, it is heated at a low red heat, when magnesia is formed, the calcium carbonate remaining almost entirely unaltered, since it requires nearly a white heat for its decomposition. The powder ground in water was, when treated with carbonic acid under a high pressure, soluble, magnesium bicarbonate being formed, the saturated solution of which contains 2·3 per cent. of magnesia, and has a specific gravity of 1·028. The solution of lead chloride contains 1 part of the salt in 126 parts of water; it is mixed with a slight excess of the magnesium carbonate solution as quickly as possible. The liquid is drawn off from the mixing vessel into a large receiver in which a precipitate deposits, consisting of white lead and a little oxychloride. After drying, the precipitate is ground with a small quantity of caustic soda to decompose the oxychloride. A few days afterwards the mass is washed to remove sodium chloride and the product dried.

The process of Dale and Milner is similar to the above magnesia process. Litharge, lead hydroxide or insoluble lead salts are mixed with sodium bicarbonate solution, and, with repeated additions of water, ground until the formation of white lead is completed. The lead compound and sodium bicarbonate are used in equivalent proportions.

According to the process of P. Bronner (German patent 52,262), 3 molecules of freshly-precipitated lead sulphate are heated with a solution of 2 molecules of caustic soda, when the basic sulphate $2\,PbSO_4.Pb(OH)_2$ is formed according to the equation—

$$3\,PbSO_4 + 2\,NaOH = 2\,PbSO_4.Pb(OH)_2 + Na_2SO_4.$$

Or 4 molecules of lead sulphate are decomposed by 2 molecules of caustic soda, according to the equation—

$$4\,PbSO_4 + 2\,NaOH = 3\,PbSO_4.Pb(OH)_2 + Na_2SO_4.$$

This transformation takes place at a temperature of 70° C. The resulting basic sulphate, although pure white, cannot be used as a pigment on account of its lack of covering power; but by heating with a solution of sodium carbonate it is converted into white lead.

$$2\,PbSO_4.Pb(OH)_2 + 2\,Na_2CO_3 = 2\,PbCO_3.Pb(OH)_2 + 2\,Na_2SO_4.$$
$$3\,PbSO_4.Pb(OH)_2 + 3\,Na_2CO_3 = 3\,PbCO_3.Pb(OH)_2 + 3\,Na_2SO_4.$$

By this process, which is harmless to the workmen, the lead sulphate obtained as a by-product in the preparation of mordants for calico printing, can be converted into good saleable white lead. The lead sulphate may also be obtained from litharge, lead acetate or nitrate.

It occasionally happens that white lead has a rose tint, which is clearly perceptible by comparison with a pure white sample. This colouration occurs in white lead made from argentiferous lead. A very small quantity of silver is sufficient to produce the tinge of colour.

Occasionally white lead which has been ground in oil and used for painting turns perceptibly yellow, the colouration being similar to that observed on a surface painted with white lead from which light is almost excluded. The yellow colouration is due to lead oxide. This has been proved by suspending such a white lead in water and treating it with carbonic acid, after which a surface painted with it remains permanently white.

(f) Oxychloride White Lead.

Under the name of white lead, but differing from it in composition, various products are found which consist of lead oxychloride. This compound is also known as Pattison's white lead.

Pattison's white lead can be much more cheaply manufactured than real white lead, the raw material employed being the cheap galena. The finely-powdered mineral is boiled with strong hydrochloric acid in closed lead vessels. Sulphuretted hydrogen is evolved, which may be burnt to sulphur dioxide and so used to make sulphuric acid. A hot saturated solution of lead chloride remains, from which the salt separates in small crystals on cooling. The crystals are drained in a basket and washed with cold water to remove the acid. The pure lead chloride is then dissolved in hot water and mixed with lime water. Pattison obtained lime water from dolomite by burning it, treating with a little water to remove the easily soluble salts, and, after the removal of this wash water, treating the residue repeatedly with water in order to obtain a clear solution of pure hydrate of lime. When pure limestone is used, it may be treated with water immediately after burning without any preliminary preparation.

Two equivalents of lead chloride are used to one equivalent of calcium hydroxide. Practical experience showed that the best product was obtained when the precipitation was very rapidly brought about. With this object, both solutions entered the precipitation tanks through pipes with narrow slits at the side, so that the liquids met in a thin layer, in which the precipitation of the pigment was instantaneous. It is also necessary that lead chloride should be in excess throughout. The liquid is allowed to stand for the precipitate to settle, which it does in a brief time on account of its high specific gravity. The solution now contains the small excess of lead chloride in addition to calcium chloride; lime water is added until the liquid turns red litmus paper blue. From the alkaline solution all the lead soon separates as lead hydroxide, which is dissolved in hydrochloric acid, and thus again comes into the process.

In order to utilise the large quantities of hydrochloric acid obtained in the manufacture of soda, Percy described a process in which galena is ground with hydrochloric acid, whereby in 30 to 40 hours all the lead is converted into lead chloride, whilst the stony admixtures are unattacked. The lead chloride is then separated by levigation from the undissolved minerals and washed until free from iron, when it is dissolved in hot water and converted into oxychloride by means of lime water.

Lead Sulphite, $PbSO_3$, can be obtained by passing sulphur dioxide into a solution of basic lead acetate; lead sulphite is precipitated and a solution of neutral lead acetate remains. The process is similar to the French white lead process, with the difference that sulphur dioxide is used instead of carbon dioxide. Lead sulphite has no advantages over white lead, and is more expensive; it has thus never found practical application.

Lewis and Bartlett's White Lead Pigment.—In the lead works at Zoplin, in Missouri, galena is smelted with limestone and coal, lead fume being obtained in addition to metallic lead. The lead fume deposits are ignited, and again worked for lead and lead fume. This last lead fume can at once be used as a white pigment; it consists principally of lead sulphate, lead oxide and zinc oxide.

White Lead-Antimony Pigments.

Lead antimonite and antimonate are both heavy, white powders which can be used as pigments. They are dearer than white lead, to which they are inferior in covering power, and which they do not exceed in permanence.

Lead Antimonite is obtained by heating 5 parts of finely powdered antimony with 20 parts of sulphuric acid until a dry, white mass of antimony sulphate is left. This is fused with soda ash, the melt is extracted with water, and lead antimonite obtained by precipitating with lead acetate.

Lead Antimonate is formed by introducing in small quantities at a time a mixture of 1 part of finely powdered stibnite (antimony trisulphide) with 5 parts of sodium nitrate into a red-hot crucible, boiling the mass with water and precipitating the solution with lead acetate.

CHAPTER VIII.

ENAMEL WHITE.

BARIUM sulphate, known as permanent white, enamel white, *blanc fixe*, barytes white, is the only white pigment which is absolutely unaltered by exposure to the atmosphere. Lead pigments are discoloured in the course of time, and in the end turn black; bismuth white behaves in the same manner; zinc white is much more lasting, but not quite permanent.

Enamel white is really permanent; it deserves the greatest attention from the colour manufacturer, especially as it can be made by a very simple and cheap method. When sulphuric acid or a soluble sulphate is added to the solution of a barium salt, all the barium is at once precipitated in the form of barium sulphate.

When quite pure, barium sulphate forms an extremely soft, brilliantly white powder, which offers complete resistance to the action of the atmosphere, and also of strong acids and alkalis. It is extremely insoluble, and can be precipitated from the most dilute solutions, and is then obtained in so fine a state of division that it cannot be filtered from the liquid; it passes through the closest filter together with the liquid. When the barium solution is heated to boiling before precipitation, the precipitate is somewhat coarser, and can be filtered off without difficulty.

Barium sulphate occurs ready formed in nature as the mineral barytes or heavy spar. Finely ground barytes may be used alone as a pigment, but more commonly is used for reducing white lead, for which purpose it is particularly applicable on account of its high specific gravity. This admixture must be regarded as diminishing the quality of the pigment, because ground barytes has far less covering power than white lead. Artificial barium sulphate is in a state of division which cannot be reached by grinding barytes, consequently it considerably surpasses the latter in covering power.

The raw material for the manufacture of enamel white is either barytes or witherite (barium carbonate); the latter, however, occurs so rarely, in comparison with barytes, that the greater quantity of all barium compounds is obtained from barytes.

If witherite is obtainable in large quantity, enamel white can be prepared from it by dissolving in hydrochloric acid and precipitating the solution of barium chloride so obtained by sulphuric acid. If the witherite is very pure, the process may be simplified by treating the mineral directly with sulphuric acid, and separating the enamel white by a process of levigation from the impurities. In this case it is, however, necessary to add a small quantity of hydrochloric acid to the sulphuric acid, for the latter forms on the surface of the witherite, at the commencement of the reaction, a thin layer of barium sulphate, which is quite sufficient to prevent the further action of the acid on the witherite lying below. The hydrochloric acid forms barium chloride, which is at once decomposed by the sulphuric acid into barium sulphate and free hydrochloric acid; this again dissolves a fresh quantity of witherite, and this process is repeated until the mineral is completely and quickly dissolved.

Enamel white is, however, generally prepared from barytes, which is ground into a very fine powder and converted into barium sulphide by heating with coal (see pages

41 and 42). Hydrochloric acid acting on the sulphide produces barium chloride and sulphuretted hydrogen.

The covering power of a pigment is greater the finer its state of division, so that it would appear advisable to precipitate a weak solution of barium chloride by sulphuric acid at the ordinary temperature. When the barium sulphate has been completely precipitated, a solution of pure hydrochloric acid remains, which ought to be utilised; but when very dilute barium chloride solution is used, the hydrochloric acid is so dilute as to be useless. The barium chloride is, therefore, given in practice such a strength that it has a specific gravity of about 1·198; when the barium sulphate has been precipitated from this solution the residual hydrochloric acid has a specific gravity of 1·043.

Water of considerable purity must be used to dissolve the barium chloride. Experience has shown that water which contains appreciable quantities of organic matter does not give a pure white product. The presence of sulphate of lime in the water, which precipitates barium sulphate, need not be regarded, because the barium sulphate is so finely divided that it remains suspended in the liquid, and is carried down on precipitation of the enamel white by sulphuric acid. Carbonate of lime in the water causes the separation of barium carbonate; this may be avoided by slightly acidifying the barium chloride solution, thus converting the calcium carbonate into chloride.

According to C. A. F. Meissner, artificial barytes, suitable for use in oil paints, is obtained by precipitating barium salts by soluble sulphates in place of sulphuric acid, then quickly heating the washed and dried precipitate in a muffle to a red heat and throwing into cold water.

As has been already stated, enamel white is the most permanent pigment that exists; it appears destined in course of time to replace white lead and all other white pigments, especially as its cost is generally lower than that of the other white pigments. It costs, for example, only half as much as white lead. At present, the principal uses of enamel white are found in paper staining; it is not used to any extent in oil paints. On account of its permanence, it should be used in the place of white lead and white zinc. It also appears particularly suitable for obtaining pale shades; it can be mixed with any other pigment in any quantity without altering it in the least. This is, of course, only true when the enamel white is completely pure, and when it has been freed from every trace of hydrochloric acid by careful washing.

Lithopone.

A white pigment is obtained, according to Orr's process, by lixiviating crude barium sulphide, obtained by igniting barytes with coal, with water and dividing the solution into two parts. Zinc chloride is added to the first portion, then zinc sulphate, and finally the second portion of barium sulphide solution. The white precipitate obtained by this process contains one equivalent of barium sulphate to two equivalents of zinc sulphide. It is collected, quickly dried, heated in retorts to redness, and, whilst still hot, thrown into cold water, by which its density and therefore covering power are increased. The pigment is finally washed and ground. It is a good white, but when mixed with lead pigments discolours them by reason of the sulphide it contains.

CHAPTER IX.

WASHING APPARATUS.

IN all colour works operating on a large scale, special apparatus is used, in which are carried out the washing, pressing and drying of the pigments obtained as precipitates. Only pigments prepared in small quantities are filtered through filter paper. The treatment of enamel white and white lead requires the use of apparatus for this purpose in a special degree. We insert a short description here.

The preparation of enamel white takes place most conveniently in tubs provided with a stirring apparatus. When the precipitate of barium sulphate has once settled to the bottom, it is very difficult to again mix it up with water by means of a hand stirrer, an operation which must be often performed in washing. If vessels be used provided with a suitable mechanical stirrer, the precipitates are rapidly and thoroughly washed. We have already stated that vessels with a stirrer capable of being raised out of the liquid were specially suitable for washing white lead. Such an arrangement can with advantage be used in washing all precipitates. Many of the mineral pigments have to be freed from admixed salts by washing. A description follows of an effective washing apparatus with movable stirrer. In the cylindrical vessel (Fig. 21), which may be of any size, is a vertical iron shaft rotating upon a pin in the bottom of the vessel. On this axis is a horizontal wooden crosspiece, the under surface of which is studded with brushes; the disc, which this crosspiece carries, is united by means of two bars with a second, in which is cut a screw moving on the screwed shaft. The shaft is rotated by means of the cogwheels shown in the illustration. The handle fastened to the upper disc enables the crosspiece carrying the brushes to be raised or lowered. When the handle is held fast, this crosspiece rises or falls according to the direction of rotation of the axis. The pipe shown at the side supplies the water for washing the precipitates.

In the preparation of colours liquids have often to be

Fig. 21.

brought into the precipitating vessel which would attack the iron of the stirring apparatus, so that it is advisable to make the connecting rods between the two discs of such a length that the screws may be above the vessel. All iron parts of the apparatus dipping into the liquid should be protected by asphaltum varnish.

When a specially heavy precipitate is to be washed, such as enamel white, chrome yellow or white lead, the stirrer is raised as high as possible before the commencement of the operation. When the precipitate has formed and settled, the liquid is run off, the water tap opened, and the stirrer slowly brought down to the precipitate; the brushes fastened to the crosspiece stir up first the top of the precipitate, then the next portions, and so on until the whole of the precipitate has been stirred up into the liquid. When this has been accomplished, the stirrer is kept going for some time, so that the water may take up as much of the soluble materials as possible; it is then raised out of the vessel, in which the precipitate again settles.

As a rule, two or three washings of permanent white, in an apparatus of the construction described, are sufficient to render it quite free from acid. Washing must be continued until the wash water leaves blue litmus paper quite unchanged. When dry, precipitated barium sulphate is a very soft powder of great whiteness, which, on account of the fineness of its particles, can be readily ground with binding materials.

Enamel white loses, in a remarkable manner when completely dry, a great portion of its covering power and of its valuable property of being easily ground with oil or size to a homogeneous paste. It is not known whether this alteration is caused by a molecular change of the barium sulphate, as is not altogether improbable. In order to preserve the valuable properties of enamel white it is not, as a rule, completely dried, but is brought into the market in the form of pulp, which is obtained by bringing the washed precipitate into strong linen bags and allowing the water to drop through. This object is more quickly accomplished when the last wash water is drawn off only to such a point that, when the precipitate is again mixed up by means of the stirrer, a thin paste is formed. This paste is run into a centrifugal machine, of which Fig. 22 is an illustration. In the drum, B, provided with an outflow, R, the smaller removable drum, C, with perforated

WASHING APPARATUS.   123

walls, is caused to rotate by gearing, G, to which it is attached by the screw, V. The drum, C, is lined by a tight linen bag.

When the drum is in rapid motion, the thick liquid in the washing vessel is run in. In consequence of the centrifugal force due to the rapid rotation, the whole mass is at once thrown on the sides of the drum, the liquid penetrates the fine openings, is caught in the outer vessel, and runs

FIG. 22.

away to a receiver, in which it is kept until the finest particles of the precipitate, which will penetrate even the closest fabric, have settled. The operation is continued until the excess of water has been separated, when the bag containing the precipitate is lifted out of the drum. Centrifugalised enamel white is a fairly stiff, white paste, which should be packed in sacks lined with oiled paper to prevent drying.

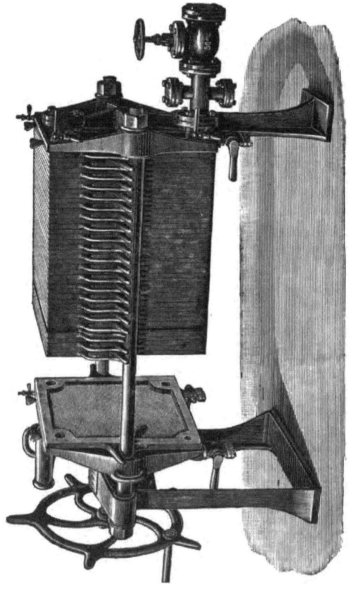

Fig. 23.

Filter Presses.

In recent years the use of filter presses for separating liquids from precipitates and for washing the latter has become general. A filter press consists, as shown in Fig. 23, of a number of frames, between which are perforated plates and sheets of filter cloth, and which can be pressed tight together by a screw. A powerful pump forces the liquid containing the precipitate into the hollow spaces of the frames (known as chambers), in which the solid body remains, whilst the liquid goes through the filter cloth. By afterwards pumping fresh water through, the solid remaining in the chambers is soon completely washed.

CHAPTER X.

ZINC WHITE.

ZINC OXIDE, known as a pigment under the name of zinc white, is one of the most important white pigments. Although not absolutely permanent, yet it has, in common with enamel white, the valuable property of preserving its whiteness in air containing sulphuretted hydrogen. Its low price has brought it into general use.

Although zinc white is a most important pigment, it is very seldom made in colour works, because on account of its origin it is a product of metallurgical processes. Zinc white in chemical composition is pure zinc oxide; it is formed when zinc vapour burns in air. In zinc-smelting works zinc white is obtained by putting zinc in tubes which are heated to whiteness; the zinc vapours burn when they come in contact with the air, and the zinc oxide is caught by special arrangements.

The retorts used for this purpose are similar to those employed in the manufacture of coal gas. From 8 to 18 of these retorts are arranged in a furnace, in two rows, one above the other. In the lid of the retorts is an opening, which serves for charging and for carrying off the zinc vapours. When the operation is commenced, the retorts being heated to a white heat, two zinc plates are brought into each, the metal is soon volatilised, and the vapours pass through the above-mentioned openings. A current of air, heated to 300° C., is blown in to meet the zinc vapours, which take fire and burn with a blinding white flame, producing a very fine white powder, which is carried by the current of air through a series of chambers in which it deposits.

C. Freitag recommends for the production of zinc white the use of retorts of oval section, A, Fig. 24. These retorts, containing crude zinc, are heated to a white heat, and then a mixture of generator gas from coke and air is introduced by B and the pipe running through B. The zinc burns completely in the flame of the generator gas, which contains

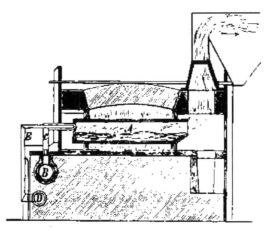

FIG. 24.

excess of oxygen. A product of faultless nature is said to be obtained in this way.

In zinc works zinc white is always made in the manner which has been described. It may also be obtained as a by-product in another metallurgical operation, the desilverisation of lead by Parkes' process. In this process an alloy of silver and zinc is obtained. By sending a current of superheated steam over the molten alloy the zinc decomposes the steam, hydrogen and zinc oxide being formed. The zinc white is carried by the current of gas into chambers, in which it deposits.

The zinc white obtained by burning zinc is, as has been said, a very fine pure white powder, which can at once be used for paint without further preparation. The price of zinc white is rather higher than that of white lead, but the difference is counterbalanced by the greater covering power of the zinc white. Ten parts by weight of zinc white completely cover a surface for which 13 parts of white lead are required.

Whilst white lead cannot be mixed with many pigments, such as those which contain sulphur, zinc white may be mixed with all without fear of alteration. Zinc white is even better than enamel white for producing pale pigments from lakes; it has a lower specific gravity than enamel white, so that the mixture with the light lake can be more easily made.

Zinc grey, which is produced by some works, is zinc oxide discoloured by metallic zinc. Pure zinc white always has a pure white colour; if it is tinged with grey it is contaminated by metallic zinc, whilst a brownish hue denotes the presence of cadmium oxide. The latter impurity will be rarely met with in commercial samples, since cadmium is worth much more than zinc. Zinc oxide is used by the colour maker in the preparation of Rinmann's green; it is also used, as stated above, as an addition to other pigments.

Griffith's Zinc White.—This pigment, which is equal in covering power to white lead, consists of zinc oxysulphide. It is obtained by precipitating a zinc solution with a solution of barium sulphide, washing, drying, igniting and grinding the precipitate. As it contains sulphur, it should not be mixed with copper or lead pigments.

Tungsten White (Lead Tungstate), $PbWO_4$, is obtained as a heavy powder by precipitating a solution of sodium tungstate with lead acetate and treating the precipitate, which consists of basic lead tungstate, with dilute acetic acid, by which lead oxide is dissolved and a salt of the above composition left. This white pigment is dearer than other lead pigments, and has no special advantages over them; it is, therefore, seldom used.

White Antimony Pigments.

Antimony forms a number of white compounds which can be made by a simple and inexpensive process, and might, therefore, be used as pigments. Two antimony compounds in particular are so used—antimony trioxide and oxychloride (powder of algaroth).

Antimony Trioxide occurs ready formed in nature as white antimony or antimony bloom. It is very simply prepared by burning the metal in air, when it forms soft needles with a silvery lustre. It is only necessary to heat melted antimony to a little above its melting point in a crucible placed in a slanting position, when the metal takes fire and burns with a blue flame. Nitric acid converts metallic antimony very quickly into antimony oxide with a copious evolution of brown fumes.

Antimony trioxide may be more cheaply prepared from antimony sulphide, artificial or natural (stibnite), by finely powdering, moistening with water, and gently warming on plates. Oxidation takes place, the sulphur is converted into sulphur dioxide and the antimony to trioxide. The heating should not be carried too far, or the antimony takes up further oxygen and forms the tetroxide.

Antimony Oxychloride, or Powder of Algaroth, is obtained by dissolving stibnite in strong hydrochloric acid, the operation being performed under a chimney to carry off the sulphuretted hydrogen. The solution of antimony trichloride, when the impurities have settled, is poured into a large vessel of water. At once a pure white precipitate is formed, which quickly sinks; it is washed with water until the washings have no acid reaction. Washing should not be continued too long or the oxychloride will be further decomposed; by washing with hot water it is almost entirely changed to antimony trioxide. The precipitate, after washing with cold water, has generally the composition expressed by the formula SbOCl.

Antimony oxide and oxychloride are both very crystalline powders, and have in consequence small covering power.

Bismuth White is not used as a painters' pigment; it has no advantage over the white pigments previously described, and is much more expensive. Its only use as a pigment is for the preparation of white cosmetics, and even for this purpose zinc white is now frequently used; it is cheaper and quite as satisfactory.

Bismuth white is prepared by treating metallic bismuth with fuming nitric acid. The white precipitate at first formed completely dissolves in an excess of acid, and when the solution is poured into a large quantity of water, basic bismuth nitrate—bismuth white—separates.

Pure bismuth white is a soft, heavy powder, brilliantly white; it must be preserved in air-tight vessels as soon as it is dry, otherwise it acquires a yellowish tinge. Bismuth compounds are, if possible, more susceptible to the action of sulphuretted hydrogen than lead compounds. The yellow colouration, turning to black in the course of time, is due to black bismuth sulphide.

Tin White is used for earthenware enamels. It is obtained by treating granulated metallic tin with very strong fuming nitric acid. The heavy, white powder which is formed is separated from the undissolved tin by floating. Tin white has no application as a pigment; when mixed with glazes it gives them a handsome, milky appearance.

Manganese White.—When large quantities of a solution of impure manganese chloride are at hand, such as are produced in the preparation of chlorine, manganese carbonate may be obtained. A small quantity of soda solution is first added and the liquid left for several days so that the oxide of iron may separate. When this is the case an addition of soda gives a pure white precipitate.

Magnesia White or Mineral White is obtained, according to T. H. Cobley, by mixing a solution of magnesium sulphate with calcium chloride, adding 10 per cent. of aluminium chloride to the mixture and stirring in slaked lime so long as a precipitate is formed. A cheaper process is to precipitate mixed solutions of magnesium and aluminium sulphates by slaked lime.

Annaline.—A white pigment is recommended for use under this name; it consists of dead-burnt gypsum, which has been converted into a fine powder by grinding and levigating. (Dead-burnt gypsum has been so strongly heated that it is not able to again unite with water.)

To obtain paler shades of certain colours, additions are made of natural pigments, such as chalk, which has been converted into a very fine powder by levigation. An addition of chalk to a heavy mineral pigment, such as chrome yellow, is not advisable; it would, besides, seriously diminish its covering power.

In the preceding pages, a large number of white pigments has been enumerated. It would be easily possible to increase the list, but the result would be of no practical interest, for other pigments, neither in respect of quality nor price, can compete with the cheaper white pigments. Although white lead has at the present time an enormous use, it is to be hoped that this pigment, of good colour but little permanence, may be replaced entirely in the course of time by zinc white, and for some purposes by enamel white.

White pigments, in addition to their use alone, are employed to produce tints, which are obtained by mixing deeper colours with the white pigment. By adding the proper quantity of white to a colour it is possible to produce all paler shades of that colour. For example, the different varieties of the red lakes which are found in commerce are obtained by mixing white pigments with the deep red lakes.

The particular white pigment to be employed in these mixtures depends on the nature of the colour and on its specific gravity. It should always be remembered that white lead will not increase the permanence of a colour, since it will be discoloured in a short time by the action of the atmosphere. In the manufacture of fine colours for artists lead pigments should be absolutely excluded.

To produce paler shades of colours which contain lead and therefore have a high specific gravity, *e.g.*, chrome yellow, white lead may be used; for other colours, barytes or zinc white should be employed. In the case of lakes enamel white is too heavy; zinc white or magnesia is suitable.

CHAPTER XI.

YELLOW MINERAL PIGMENTS.

As was the case with white pigments, so with yellow: of the large number known very few are in extensive use. In former times the number of yellow pigments employed in painting was far greater than at present; several, formerly in general use, have dropped out, partially or entirely, owing either to their poisonous character or to their replacement by others, deeper and more handsome. Especially since the discovery of cadmium yellow and the development of the manufacture of chrome yellows, many colours once in general use have properly fallen into disuse.

Again, unfortunately, the most important of the yellow mineral pigments contain lead, and have little stability; but a series of yellow colours free from lead is known, and though some of them are inferior in shade to the lead pigments they surpass the latter in permanence.

In addition to lead compounds, pigments derived from barium, zinc, antimony and cadmium are in general use. The yellow lead pigments were formerly preferred, and at present, so far as concerns beauty, they must be regarded as superior to other mineral yellows. The endeavour to provide the artist with permanent colours has resulted in the use of others, perhaps less brilliant, but very lasting. Under the name of chrome yellow a single pigment consisting of lead chromate was formerly understood; to-day, under this title is comprised a series of pigments containing zinc or barium in place of lead, but all commercially known as chrome yellow.

CHAPTER XII.

CHROME YELLOWS.

THE chrome yellows are the lead, zinc or barium salts of chromic acid. In describing the different methods used in making these colours, lead chrome yellow will be taken first, since this is the one most largely used.

LEAD CHROME YELLOW.

Both neutral and basic lead salts of chromic acid are known. Neutral lead chromate is found in nature as the somewhat rare mineral crocoisite, which is found in very small but perfectly shaped crystals in many lead mines.

Neutral Lead Chromate, $PbCrO_4$, is formed as a very heavy precipitate of a fine deep yellow colour when a solution of potassium chromate or bichromate is added to a solution of a lead salt in water. When exactly equivalent quantities of potassium chromate and lead solution are used, and the strength of the solutions is the same, the product has the same shade each time the operation is performed. It is not immaterial whether the one or the other salt is in excess, or whether strong or weak solutions are used; all these conditions modify the shade of the chrome yellow produced. Many colour makers are apparently of the opinion that some particular skill of the workman is necessary to produce chrome yellow of a particular shade. This is, however, not the case; manufacturers who know the simple conditions which are important in making chrome yellows may produce any desired shade without difficulty.

Neutral lead chromate readily parts with half the chromic acid it contains. When treated with alkalis, such as lime or caustic soda, or even when digested with finely ground litharge, the neutral salt gives up half of its chromic acid, and is converted into the basic chromate or chrome red, Pb_2CrO_5.

Basic lead chromate has, as the name chrome red indicates, a fine red colour. If the quantity of lime or caustic soda used is sufficient to decompose only a portion of the neutral lead chromate, a mixture of the yellow neutral and the red basic chromate is formed, the shade of which will incline to yellow or red according as it contains a preponderance of one or the other compound. The pigment known as orange chrome is a mixture of approximately equal parts of the neutral and basic lead chromates.

In order to brighten the deep yellow shade which distinguishes neutral lead chromate, it is either mixed with a white pigment, or a white substance (lead sulphate) is precipitated from the solution simultaneously with the lead chromate. In this manner all the imaginable pale yellow shades, lemon yellow, sulphur yellow, etc., can be obtained.

Just as the quantities of the solutions used and their strength influence the shade of the chrome yellow they produce, so it also appears to be not immaterial which lead salt is employed. Colour makers are generally agreed that the finest product is obtained from neutral lead acetate. Any lead salt, even insoluble in water, may be used for the preparation of chrome yellow. The affinity of chromic acid for lead is so great that an interchange of constituents occurs between the insoluble salt and the potassium chromate. Chrome yellow may be made from lead acetate, chloride or sulphate; the resulting substances are the same, but there is a considerable difference in regard to the beauty of the product. The finest chrome yellow, which leaves nothing to be desired in beauty of shade, is obtained by proceeding in the following manner : Lead acetate is dissolved in water, the solution diluted with an equal volume of water, and then mixed, under constant stirring, with a similarly diluted solution of potassium chromate or bichromate. The precipitate is immediately formed, and quickly sinks to the bottom in consequence of its high specific gravity. It is washed with clean water so long as this removes soluble salts. The precipitate is then drained on cloths and dried in the air.

The finest product is obtained by working with the following proportions :—

Sugar of lead	100
Potassium bichromate, or	50
Potassium chromate	40

If lead sulphate is used the following quantities are to be employed :—

Lead sulphate	100
Potassium bichromate	25

In the case of lead chloride the following is the proportion :—

Lead chloride	100
Potassium bichromate	27

The chrome yellows prepared from insoluble lead salts have no particular beauty, but they may be used for mixed colours such as the spurious chrome green.

Preparation of the Lead Solution. — Many makers of chrome yellow do not use commercial lead acetate, but prepare its solution themselves. The preparation of this solution requires neither much space nor labour, so that considering the high price of lead acetate this procedure may be regarded as advisable, but only when acetic acid is obtainable at a low rate.

The following is the method by which lead acetate solution is made. Lead is granulated by pouring the molten metal from a height of several yards into cold water, which is kept in rapid motion. The smaller the particles of lead, the larger surface they will possess, and the more quickly they will dissolve. For the solution of the lead small tubs are used, 50 centimetres in diameter and 90 to 100 centimetres in height, provided with a tap immediately above the bottom to run off the liquid. Four of these vessels are so placed, one above the other, that the contents of each may be run into the one next below it. The lead in the top vessel is covered with acetic acid; in a few minutes this is allowed to flow into the second vessel, and similarly after a few minutes from the second to the third, from the third to the fourth, from which it runs away into a receiver below. After this treatment it contains but a small quantity of lead acetate; the object of the operation is to start the oxidation of the lead, which quickly follows when air has sufficient access, the lead particles lose their metallic appearance and become covered with a white layer. When this is the case, the acetic acid is pumped from the receiver back into the top vessel, where it is left one to two hours in contact with the lead. It is then run off into the second, and thence into the third and fourth, remaining in each vessel for about the same time; the resulting liquid is an almost completely neutral solution of lead acetate. The treatment with acetic acid is continued so long as lead remains undissolved.

The solution of potassium chromate is prepared in a tub. The salt is easily soluble, and warming is unnecessary if the chromate is placed in a basket, lined with close linen cloth, hung in the liquid so that it is immersed to half its depth. The salt rapidly dissolves, its solution has a greater density than water, in consequence of which it sinks and fresh quantities of water continually come in contact with the salt.

Precipitation of the Chrome Yellow.—Before the chrome yellow can be precipitated it is necessary to estimate the quantity of lead acetate contained in the solution, since upon this depends the quantity of potassium chromate solution to be used. If the lead solution contained only acetate and water, its strength could be simply found by means of the hydrometer. It contains, however, varying quantities of acetic acid, on which account the hydrometer would give very inaccurate results. The test by which the relation between lead solution and potassium chromate solution is found is performed in the following manner: The lead solution is measured off in a cylinder divided into 100 divisions; the same volume of potassium chromate solution is measured and placed in a high narrow vessel; the lead solution is gradually added to the chromate solution so long as a precipitate is formed. The precipitate settles rapidly, and there is no particular difficulty with a little practice in finding with sufficient accuracy the quantities required for the precipitation. In order to precipitate 100 litres of potassium chromate solution, there are required as many litres of lead solution as were used divisions of the cylinder.

The preparation of the chrome yellow is now a very simple matter. Whilst steadily stirring, the measured quantity of lead solution is run into the solution of potassium chromate; the precipitate is allowed to settle, is well washed and dried. It does not make any difference to the colour whether potassium chromate or bichromate is used; the same product is obtained in each case.

It is stated by Dullo that chrome yellow prepared by the preceding process alters in colour on long keeping. This is ascribed to the formation of a basic compound. According to the same author, a chrome yellow free from this objectionable property is obtained by using lead nitrate in place of acetate and an excess of potassium chromate solution. The writer has kept chrome yellow, made from lead acetate, for years without observing the slightest alteration in the colour.

On chemical grounds, it is incomprehensible that a chrome yellow prepared from lead nitrate should have different properties to the same substance prepared from another soluble lead salt, and freed from foreign substances by sufficient washing.

The product of this process is that which *par excellence* is known as chrome yellow, the chemist's neutral lead chromate; it exhibits a characteristic deep yellow colour, a shade which is known as chrome yellow. Under the microscope chrome yellow is seen to be a crystalline mass; it will possess greater covering power the smaller the crystals. Now, the motion of a liquid in which crystals are forming prevents the production of large crystals, thus the reason is clear for the rapid stirring of the solutions in the preparation of this pigment.

According to C. O. Weber, who has published an exhaustive account of chrome pigments (*Dingler's Journal*, 282), the cost of the lead chrome pigments varies greatly according to the raw materials employed. Assuming that 100 kilogrammes of litharge cost 35 marks, 100 kilogrammes of 30 per cent. acetic acid 25 marks, and 100 kilogrammes of 60 per cent. nitric acid 26 marks, Weber calculates that 100 kilogrammes of litharge, in a form suitable for making chrome yellow, will cost as follows:—

From lead acetate at 56 marks per 100 kilogrammes	96 marks
,, ,, solution made in the works	80 ,,
,, lead nitrate at 50 marks per 100 kilogrammes	75 ,,
,, ,, solution made in the works	64 ,,
By the basic lead acetate method	51 ,,
,, ,, chloride ,,	40 ,,
White lead method	55 ,,

The Pale Chrome Yellows.—When the solution of the chromate used for the precipitation of the lead solution is mixed with sulphuric acid, then a mixture of lead sulphate and lead chromate is formed on precipitation. Lead sulphate is white, so that the colour of the precipitate would be paler according to the quantity of sulphuric acid added to the potassium chromate solution. There are, however, compounds of lead chromate and lead sulphate, of which we know two. Their composition is expressed by the formulæ:—

$$PbCrO_4 . PbSO_4 \text{ and } PbCrO_4 . 2PbSO_4.$$

The former is a beautiful lemon yellow shade, the latter nearly approaches sulphur yellow. By corresponding alterations in the quantity of sulphuric acid added, all intermediate shades can be obtained. On the works these shades are made in the following manner: buckets are used for taking the potassium chromate solution out of the vessel in which it was made; these buckets hold 12·5 litres. Now, if the solution of potassium chromate has been made from 25 kilogrammes of the salt and 750 kilogrammes of water, one of these buckets holds exactly 0·43 kilogramme of chromic acid. In order to obtain the lemon yellow compound, 0·39 kilogramme of sulphuric acid must be added to a bucketful of solution; 0·78 kilogramme must be added to obtain the sulphur yellow chrome. These liquids are prepared by pouring the sulphuric acid in a thin stream into the potassium chromate solution made as above, the liquid must be stirred whilst the sulphuric acid is being added. The lemon chrome has the peculiar property of increasing considerably in volume soon after formation, a property to which regard must be had in the manufacture. A description of the rational preparation of the lemon and sulphur yellow shades of chrome yellow on the manufacturing scale follows.

In making lemon chrome, the tub in which the precipitation is to take place is two thirds filled with water, the lead solution is stirred in, and then the chromate solution, mixed with the proper quantity of sulphuric acid, is run in; the liquid is well stirred so that the precipitate may form as quickly as possible throughout the whole liquid. The precipitate is allowed to settle, the liquid drawn off, and the colour washed twice with water as quickly as possible. The paste is then removed from the tub and poured on a strong linen strainer. At first, the fine precipitate goes through the strainer, the liquid is poured back on the strainer until the size of its pores is so far diminished by the precipitate itself that only clear liquid runs through. The precipitate is left on the strainer until it forms a stiff mass, which can be easily spread out upon boards by spatulas.

To obtain a good, that is, a loose product, it is necessary to carry out the processes so quickly that the swelling mentioned above does not take place while the colour is being strained, but when it has been spread out on the boards. This swelling only takes place completely when the layer of precipitate is fairly thin. Large boards should be used, upon which the precipitate is spread out in a very thin layer. To prevent the mass—which is still fairly fluid—from running off the boards, they are provided with raised edges, and the paste is spread out smoothly in these flat trays. If the operations have been properly performed, the precipitate at once begins to swell and becomes of a loose nature. When it has acquired a buttery consistency it is cut up, by means of a thin sheet of brass, into prisms, which are placed near one another standing on the narrow side, and dried first in the shade and then in the sun. It is necessary that the drying should take place slowly at first, or the cakes will crack or even fall to pieces. The precipitate cannot be washed completely in the tub, because it often begins to swell on the strainers, consequently a crystalline crust covers the surface of the cakes during drying. This layer must be removed by scraping the cakes of colour, in which operation small quantities of chrome yellow dust become suspended in the air. To protect the workman against poisoning by this lead compound, precautions must be taken against breathing in the dust. The simplest and most efficacious is to tie a wet sponge over mouth and nose; this retains the particles of dust in the inspired air. The dust, scraped off the cakes, is put into water, in which the salts dissolve, whilst the chrome yellow sinks to the bottom. In this process for preparing chrome yellow the solution of potassium acetate left after the precipitation contains a considerable quantity of free acetic acid, which may be utilised to dissolve lead.

When sulphur yellow chrome is to be made, the process is substantially the same as for the preparation of the lemon shade, but with the difference that everything is done to prevent the swelling of the precipitate. The precipitation and washing of the precipitate are done as quickly as possible: the washed precipitate is filled into press bags and strongly pressed. Care must be taken in the pressing that the pressure is only gradually increased; if powerful pressure is applied at once, even the strongest cloths will be burst. The more thoroughly the precipitate is pressed, the closer will be the fracture of the chrome, a property which is regarded as a sign of good quality in this species of chrome yellow.

The manufacture of chrome yellow is intimately connected with that of a number of colours, varying from orange to dark red, which are known under the names of chrome orange or chrome red. In accordance with the division adopted of pigments according to their colour, chrome orange and chrome red will be considered among the red mineral pigments.

CHAPTER XIII.

LEAD OXIDE PIGMENTS.

LEAD MONOXIDE, PbO, exists in two different modifications; in the crystalline form, as litharge, which is generally pale yellow with a reddish tinge, and amorphous, as massicot, which is yellowish red.

Lead monoxide is a product of metallurgical works rather than of colour works; still the preparation of the crude oxide falls in the domain of the colour maker, since from litharge several pigments may be prepared by a simple treatment.

Massicot is obtained by heating white lead, lead nitrate or red lead, and also by heating melted lead in the air, with the precaution that the oxide formed does not itself melt.

Litharge is obtained as a by-product in several metallurgical processes, such as the cupellation of lead containing silver, in which the lead is melted in a furnace with a shallow hearth and a powerful current of air blown over the melted metal. The lead is oxidised, the oxide melts at the high temperature approaching $1000°$ C., and flows through an orifice in the wall of the furnace, whilst the silver remains on the cupel. The litharge is ground and levigated and, according as it is pale yellow or reddish, sold under the name of silver or gold litharge.

Both massicot and litharge have no particularly striking colour; they are seldom used as pigments. Litharge has an extensive use in oil boiling and for the manufacture of lead peroxide, which is used for matches.

Red Lead, Minium. — Lead forms a number of other oxygen compounds, one of which, red lead, has the composition, Pb_3O_4. It is a bright red powder, used as a pigment and as a constituent of certain cements (for gas and water pipes).

Red lead is, like litharge and massicot, a metallurgical product, but, by working on a small scale, products can be obtained of a much brighter colour than the produce of the large scale.

Red lead is made in two ways—directly from metallic

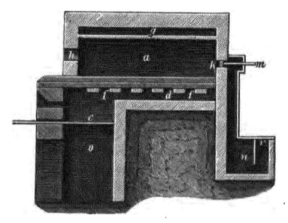

Fig. 25.

lead or by heating easily decomposed lead salts. When it is made from the metal, the following is the process: the lead, which must be very pure, is melted in a reverberatory or calcining furnace, oxidised to massicot by the air passing over it, and the massicot then, by careful heating, changed into red lead, a process in which particular care must be taken that the mass is not melted. By continued heating the massicot absorbs about 2 per cent. of oxygen, and changes in colour to a bright red.

The art in making red lead by this process lies in maintaining the proper temperature; the furnaces are constructed so that the temperature may be regulated during the heating. A reverberatory furnace is used, in which is a stirring apparatus, so that the heated mass may be continually turned over to accelerate the oxidation. Muffle furnaces are also used, in which the massicot is placed in crucibles on an iron plate, which can be pulled out of the furnace for observation of the change of colour. Whatever method is used, the temperature must be so regulated that overheating of the material is avoided, otherwise litharge is formed, which is only very slowly converted into red lead.

Mercier states that the muffle furnaces are arranged as

FIG. 26.

shown in Fig. 25. The muffle, a, is 2·5 metres long and 2 metres wide; its bottom rests on an iron plate. The passage, d, running under the muffle is 20 centimetres high; it is divided by a partition, and at each end are two hearths, c, 70 centimetres long and wide. The products of combustion pass from the long passages into side channels, f, provided with dampers, go round the muffle and unite in the space, g. The flue, k, at the back of the muffle is provided with dampers, m, which exactly regulate the current of air through the muffle; n is the chamber in which are collected the particles of oxide carried over by the draught. The furnaces used for manual labour consist, according to Percy (Figs. 26 and 27) of a rectangular or circular hearth, a, of about 3 metres diameter, which is deeper in the middle and has two fireplaces, *b*. The low arch surmounting the hearth is covered with sand in order to prevent cooling.

When finished, the red lead is drawn out of the furnace and finely ground under edge-runners, or occasionally levigated. The temperature necessary in making red lead is

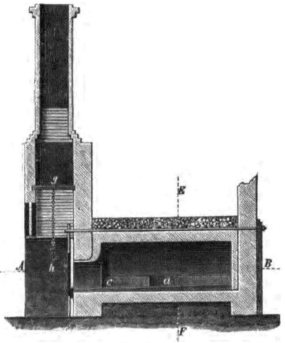

Fig. 27.

that at which the angles of the muffles, when these are used, begin to show a dark red glow.

Orange lead, which is a brighter variety of red lead, is prepared from lead salts; white lead or lead nitrite is used for this purpose. The latter salt is made by the process of Pischon, by heating 1 equivalent of lead nitrate with 4 equivalents of granulated lead and water at a temperature between 50° and 60° C. After about 2 hours, the lead nitrite separates in the form of a granular yellow mass. According to Burton's process, lead carbonate is oxidised by heating with 20 per cent. of sodium nitrate and extracting the mass with water. There are also other methods by which red lead is obtained from litharge by the use of potassium chlorate or saltpetre; but these methods, without producing a finer product than those previously given, are more expensive, and consequently have found no application on the large scale.

CHAPTER XIV.

OTHER YELLOW PIGMENTS.

Cassel Yellow, also known as mineral or Veronese yellow, has now a very restricted use; it has been replaced by the deeper and cheaper chrome yellow. Much of the Cassel yellow of commerce is nothing but chrome yellow shaded with barytes. As regards chemical composition, Cassel yellow has the following formula: $PbCl_2.7PbO$. It is obtained by heating litharge, red lead or white lead with ammonium chloride. To 10 parts of the lead compound 1 part of ammonium chloride is used; on melting, ammonia is set free, by which part of the lead oxide is decomposed, metallic lead separating. The melted mass is poured off from the lead into iron moulds, in which it solidifies to a very crystalline substance of a fine yellow colour. By grinding and levigating, the Cassel yellow is prepared for use. Pale yellow shades, obtained by admixtures of barytes, are occasionally encountered.

Montpellier Yellow consists, like the preceding pigment, of basic lead chloride. It is obtained by gradually mixing 400 parts of powdered litharge with a solution of 100 parts of common salt in 400 parts of water. After each addition of salt solution the pasty mass must be thoroughly stirred, or it will harden. When all the salt solution has been mixed with the litharge to a homogeneous white mass, the latter is treated with water to remove excess of salt, and the washed material dried and melted in earthenware crucibles.

The melt, which has a bright yellow colour, is ground and levigated, when it forms a handsome pigment.

There are several other yellow pigments of similar composition, of which one only need be mentioned, obtained by treating a solution of zinc chloride with lead hydroxide.

Turner's Yellow or English Yellow is prepared by two methods: either by melting 7 parts of finely ground litharge with 1 part of common salt; or by treating litharge with a solution of common salt and converting the white oxychloride into a yellow pigment by melting.

Naples Yellow.—This handsome pigment, which is, unfortunately, susceptible to the action of sulphuretted hydrogen, is known commercially under different names. Naples yellow takes its names from the fact that it was formerly exclusively made in Italy, where the method was kept secret, a secret which disappeared with the advance of analytical chemistry. Naples yellow is now known to be lead antimoniate.

Naples yellow is a handsome pigment. Its preparation is more tedious than that of chrome yellow, hence it is now rarely employed. The author has had practical experience that much of the so-called Naples yellow of commerce is nothing but a suitably shaded chrome yellow.

Naples yellow can be prepared by different methods. According to the oldest, given by Brunner, 1 part of pure tartar emetic is carefully and thoroughly ground with 2 parts of lead nitrate and 4 parts of common salt. The mixture is melted at a low heat in a Hessian crucible, and the fluid mass poured on a cold iron plate. After cooling, it is boiled out with water, when lead antimoniate remains as a powder of a more or less deep yellow colour. It is not easy to obtain this favourable result with certainty in every case. If a certain temperature is exceeded only by a little, a hard mass results, which by long boiling does not become a fine powder, but a sandy substance of little brilliance. Even when the operation succeeds, the product often varies considerably in shade, sometimes a sulphur yellow, at other times an orange pigment being formed. As a rule, the paler product is obtained at a lower temperature; by stronger heating, darker products of a red shade are obtained.

According to another recipe, 2 parts of tartar emetic are melted with 4 parts of lead nitrate and 8 parts of common salt. The mass is treated with very dilute hydrochloric acid for a long time, which extracts some quantity of lead oxide, a deeper product being thus obtained. Care is, however, necessary in this treatment; acid of too great strength would spoil the whole product.

The Paris method for Naples yellow is as follows: metallic antimony is oxidised by melting in air; to 12 parts of antimony, 8 parts of red lead and 4 parts of zinc oxide are used, and the mixture is melted at a low red heat.

A cheap, but not particularly bright product, can be obtained from old printer's type. The metal, which is an alloy of antimony and lead, is powdered, mixed with 3 parts of saltpetre and 4 parts of common salt, melted, and the mass washed out with water.

Other formulæ which are said to yield a good result are as follows: 12 parts of white lead, 3 parts of antimony oxide, 1 part of ammonium chloride, 1 part of alum. Or: 16 parts of stibnite, 24 parts of lead, 1 part of common salt and 1 part of ammonium chloride. The intimate mixture of these materials is first gently heated with access of air, then more strongly, and the mass extracted with water. There are many other recipes for the preparation of Naples yellow, the majority of which are distinguished by an apparently arbitrary arrangement of the materials; for there is no scientific reason. If it were possible to accurately obtain any desired high temperature in a furnace, the manufacture of Naples yellow would no longer be a matter of skill, but the same product could be obtained at every attempt. Since this is not yet the case, the exact procedure for the preparation of this colour can only be found by careful experiments.

Naples yellow is, as has been said, a handsome colour, and offers a great resistance to varied reagents. It is only changed by one of them, sulphuretted hydrogen, by the prolonged action of which it is turned completely black.

Antimony Yellow is very similar in composition to Naples yellow. It consists of a mixture of lead antimoniate with the oxides of lead and bismuth. It is prepared by the process recommended by Meromé by intimately mixing 3 parts of finely powdered bismuth with 24 parts of powdered stibnite and 64 parts of saltpetre, melting the mixture and shaking it whilst molten into water. The brittle mass is finely powdered, washed and dried, then melted with 128 parts of litharge and 8 parts of sal ammoniac. The mass obtained has a fine pale yellow colour; when powdered it is antimony yellow. This pigment has almost fallen into disuse because of its instability and the high price of bismuth.

Calcium Chrome Yellow.—Calcium forms a yellow pigment with chromic acid, which, although far surpassed by the lead chromes in fineness of shade, has the advantage over them of greater stability and cheapness. For purposes for which cheap and at the same time permanent colours are required calcium chrome yellow can be recommended. It is most simply prepared from potassium chromate and calcium chloride, which, as a by-product of many chemical operations, is obtainable at very low prices. The deepest pigment is obtained when the precipitation is done with a boiling solution of the chromate. Calcium chromate, in addition to its use alone, may be employed instead of white pigments to produce pale shades from deep lead chromes.

This addition should not be carried to a great extent or the chrome will be made too light, since calcium chromate has a much lower specific gravity than lead chromate.

Barium Yellow, Yellow Ultramarine or Permanent Yellow.—This pigment consists of barium chromate. The finest product is obtained when a solution of a barium salt, generally barium chloride, is precipitated boiling by a solution of potassium chromate. The very finely divided precipitate has a pale yellow colour very similar to that of pale lead chromes. This handsome pigment is distinguished by the valuable property of being practically unaltered by the atmosphere; it is only attacked by strong acids and alkalis. By long heating, the colour of this compound is gradually changed to a handsome green, which consists of a compound of barium and chromic oxides, and has occasional use as an artists' colour. In order to obtain this pigment, the heating must be intense and long continued. According to the author's experiments, it is not sufficient to heat for a short time to a very high temperature; in that way a mass is obtained of very unequal colour. The best result was obtained by spreading barium yellow in a thin, even layer in a flat porcelain dish and heating to whiteness for 10 hours.

Zinc Chrome Yellow.—Zinc chromate is inferior to lead chromate in beauty, but has the advantage of permanence. It does not blacken in an atmosphere of pure sulphuretted hydrogen, and resists very well the action of other agents. Zinc yellow may be prepared by the immediate precipitation of a solution of zinc sulphate by a solution of potassium chromate, both being boiling, but the very bright precipitate obtained in this way is not stable; on washing, it gives up chromic acid continually to the wash water, and only a pale yellow residue remains. A very fine colour is obtained in the following manner: zinc sulphate is dissolved in water and boiled for half an hour with 1 per cent. of white zinc whilst stirring. This operation effects the separation of iron oxide and the neutralisation of the free acid generally present in commercial zinc sulphate. When the solution has cleared by standing, it is precipitated by a solution of potassium chromate, the precipitate collected on a filter and allowed to drain completely; it is then washed with very small quantities of water and finally dried. A pure yellow precipitate is only obtained when all the iron oxide has been removed by boiling the zinc sulphate solution with white zinc; if the liquid contains only a very small quantity of iron, it has yet a very considerable influence on the colour, the yellow is not pure, but has a brownish tinge. Zinc yellow is used alone, and mixed with other pigments. Chrome yellows of all possible shades may be obtained in this way. Chrome yellows are often found in commerce which consist essentially of zinc chromate.

Cadmium Chrome Yellow.—When a solution of cadmium sulphate, or any other cadmium salt, is mixed with a solution of potassium chromate, a precipitate of cadmium chromate, $CdCrO_4$, is formed. This pigment has a beautiful, deep yellow colour, in no way inferior in shade to the finest lead chromes, and having the great advantage over the latter of being entirely unaltered by the atmosphere; it is thus to be highly recommended for artistic purposes. The high price at which it is sold prevents its general use, though now that cadmium compounds are to be obtained at so much lower prices than formerly, the price of cadmium chromate appears to be excessively high.

Cadmium Yellow is cadmium sulphide, CdS; in nature it occurs as the somewhat rare mineral greenockite. Cadmium yellow is obtained by dissolving metallic cadmium in sulphuric acid, and precipitating the solution with sulphuretted hydrogen. The solution of cadmium sulphate must be digested for some time with excess of cadmium, in order to separate the foreign metals present as impurities; the colour is not so fine when a quite pure cadmium solution is not used.

Cadmium yellow is a very bright yellow. Several shades are obtained according as the solution of cadmium sulphate used in its precipitation is neutral or acid. The reason of this difference in shade lies apparently in the different size of the crystals of which the precipitate is composed. The deep, pure yellow colour becomes still deeper by fusion, which takes place at a white heat. Weak alkalis, acids and sulphuretted hydrogen do not alter cadmium yellow; it is thus to be regarded as a durable artists' colour. It can be mixed with ultramarine without decomposition, when a fine green is formed; but mixed colours cannot be made from cadmium yellow and blue copper pigments, since these would blacken in the light.

Lead Iodide.—On precipitating a solution of lead nitrate with potassium iodide, lead iodide is formed. This is but slightly soluble in water, and, when dry, has a handsome, deep yellow colour. Unfortunately it is not permanent, but is decomposed on exposure to light. It can be used for bronzing, but other and cheaper pigments are available for this purpose.

On account of the great solubility of lead iodide in a solution of potassium iodide, it is prepared in another way, and accurately weighed quantities are used. Calcium iodide may be used instead of the potassium salt; 100 parts of iodine, 15 parts of fine iron filings and 25 parts of lime are mixed with sufficient water to form a thin paste, which is warmed until all the iodine is dissolved, when water is added, the liquid filtered, and the residue washed in order to extract all the calcium iodide. The solution and wash waters are united, then a solution of 152 parts of lead acetate is added, when all the iodine is precipitated as lead iodide.

A simpler method is to dissolve equal parts of lead nitrate and potassium iodide separately, each in 20 parts of hot water, to mix the solutions and cool quickly, when lead iodide separates in very small crystals. When pure lead iodide is melted in the absence of air, and the fused mass powdered, a product of yet finer colour is formed. It is necessary to completely imbed the crucible in which the fusion is performed in the fire. The action of air on the melted mass would produce a basic iodide. The fine golden yellow colour of lead iodide adapts it especially for the production of gold bronzes on wall papers and fabrics.

Mars Yellow, which is generally reckoned among the best artists' colours, is usually a mixture of ferric oxide and calcium sulphate or alumina. The pigment is prepared by mixing a solution of ferrous sulphate with milk of lime, when ferrous oxide is precipitated, which becomes yellowish brown on exposure to air, in consequence of the oxidation of the ferrous oxide. By heating the precipitate, according to the temperature different shades are obtained, varying between yellow and red. In addition to Mars yellow, Mars orange and Mars red are found in commerce.

The manufacture of this pigment is very simple: 1 part of ferrous sulphate is dissolved in 10 parts of water, and the solution mixed with milk of lime made from 1 part of quicklime and 40 parts of water. If it is desired to produce a darker shade, and especially a product to be afterwards converted into Mars orange, the amount of ferrous sulphate is increased to 2 parts. When the mixture has been made, it must be stirred for a long time, in order that the reacting substances may come thoroughly into contact. The precipitate, which at first is greenish grey, soon acquires by the action of the air the colour of ferric hydroxide, which becomes deeper on drying.

When dried and finely ground Mars yellow is heated in

thin layers, it changes to dark yellow, and finally to orange red, a similar alteration taking place to that occurring when ferric hydroxide itself is heated.

A Mars yellow of a deeper shade, consisting of a mixture of ferric hydroxide and alumina, is obtained by precipitating with caustic soda a solution of ferrous sulphate and alum. The sodium sulphate which is formed at the same time must be removed as completely as possible by washing with boiling water.

By calcining Mars yellow for a long time at a high temperature, Mars brown is produced, a fine brown pigment. The value of Mars yellow and the pigments obtained from it lies not only in their fine shade, but in their permanence, which distinguishes the majority of the iron colours.

Siderin Yellow.—This not very handsome yellow consists of ferric chromate; it is obtained by adding a neutral solution of ferric chloride to a strong boiling solution of potassium bichromate so long as a precipitate is formed. Siderin yellow is said to be used both in oil and water, and to be particularly adapted for use in sodium silicate paints, since in the course of time it forms a stony mass with that substance.

The low price of iron salts would make it desirable to employ chromates of iron, but it appears to be difficult to obtain a compound of constant composition. In experiments with this object the author did not succeed in obtaining products of the same shade. Others have probably been equally unsuccessful, for siderin yellow has never been used in quantity, as it would have been were there no difficulties in the way of its preparation.

Aureolin is a double nitrite of cobalt and potassium, $Co(NO_2)_2 . 3 KNO_2$. This pigment is prepared by adding excess of potassium nitrite to a solution of cobalt nitrate acidified with acetic acid. As the liquid cools, a deep lemon yellow crystalline powder separates, which, when dry, is known as Indian yellow or aureolin. It is distinguished from other yellow pigments by being unaffected by sulphuretted hydrogen.

The potassium nitrite required in the preparation of this pigment is most easily made by melting saltpetre in a thick iron vessel and stirring in fine iron filings in small quantities as soon as the saltpetre begins to decompose. The iron glows brightly and burns to oxide, the saltpetre changes to potassium nitrite. The mass is dissolved in a little hot water, the solution filtered and cooled, when most of the undecomposed saltpetre crystallises out, whilst the nitrite remains in solution. After further evaporation and separation of another crop of potassium nitrate crystals, the solution can be used to precipitate the aureolin.

It is advisable to use strong solutions in the precipitation of aureolin; the finest precipitate is obtained in this way. If dilute solutions are used, the precipitate forms gradually; it is then coarse and has little covering power.

According to the method of Hayes, aureolin is prepared by passing into a solution of cobalt nitrate the vapours produced by pouring nitric acid over copper and allowing air to enter. Caustic potash is added to the liquid from time to time. In this way all the cobalt can be obtained in the form of a yellow precipitate.

Tungsten Yellow.—Finely powdered tungsten is introduced in small quantities into fused potassium carbonate so long as effervescence occurs. After boiling with water and filtering, calcium tungstate is precipitated from the filtrate by means of calcium chloride. The moist precipitate is added to hot dilute nitric acid until the liquid is only slightly acid, when it is boiled for half an hour and allowed to cool. The precipitate, after washing with a little water and drying, is a deep lemon yellow powder.

Nickel Yellow consists of nickel phosphate. It is obtained by adding sodium phosphate to a solution of nickel sulphate or nitrate, and heating the pale green precipitate to redness. Nickel yellow has a pleasing shade, and is distinguished by great permanence. Up to the present it has found little use as an artists' colour, but on account of its permanence, which does not distinguish many yellow pigments, its use is to be recommended.

Mercury Yellow or Turpeth Mineral is a basic mercuric sulphate of the formula Hg_3SO_6. It is obtained by heating 10 parts of mercury with 15 parts of sulphuric acid in a porcelain dish in a good draught, until a white crystalline mass of neutral mercury sulphate remains. This salt, $HgSO_4$, is decomposed in contact with water into free sulphuric acid and a basic salt of the above composition. The decomposition is effected by treating the finely powdered neutral sulphate with hot water so long as the washings are acid, when a handsome lemon yellow substance remains. The wash waters contain acid mercuric sulphate. They are allowed to stand with mercuric oxide so long as this is dissolved, and the solution then used to prepare new quantities of mercury yellow.

Turpeth mineral has a very bright shade and great covering power, but it has little permanence. Sunlight soon turns it grey, and air containing sulphuretted hydrogen in a short time turns it quite black, mercury sulphide being formed.

Yellow Arsenic Pigments. — The extremely poisonous character of arsenic pigments has practically banished these handsome and cheap colours from use. In many countries their use is justly illegal. The majority of arsenic pigments have, therefore, merely historic interest. The two yellow arsenic pigments are found in nature as realgar and orpiment; though these are not rare minerals, the artificial products were generally used as pigments, and when they were in common use they were generally made in metallurgical works, in which minerals containing arsenic were treated.

Realgar, As_2S_2, has an orange red colour, whilst orpiment, As_2S_3, is a pure yellow. When these substances were used as pigments they had the same drawbacks in regard to mixing with other pigments as other sulphur compounds. King's yellow is finely powdered natural or artificial orpiment.

Lead Arsenite is a permanent deep yellow, but extremely poisonous. It can be made by fusing an intimate mixture of 100 parts of white arsenic with 75 parts of gold litharge, grinding and levigating the mass. Cadmium yellow, which has still more permanence and is less poisonous, replaces this pigment.

Thallium Pigments.—Thallium is a metal which exhibits certain similarities to lead. By precipitating a solution of a thallium salt with potassium chromate or bichromate, according to the proportion between the quantities of the two salts, precipitates are obtained of yellow, orange or deep red colour, or, after fusion, brown. By the addition of a mixture of potassium chromate and ferricyanide to a mixture of a thallium salt and ferrous sulphate an olive green pigment is obtained. On account of the rarity of thallium compounds, technical employment is out of the question, and the sensitiveness of thallium pigments towards sulphuretted hydrogen prevents their use for artistic purposes.

CHAPTER XV.

MOSAIC GOLD.

Mosaic gold consists of tin disulphide, SnS_2, fine scales of a golden yellow colour, which sublime undecomposed at a fairly high temperature and withstand the action of chemical reagents. It has a peculiar greasy nature, and can be easily ground. It is, therefore, much used for bronzing picture frames, as a pigment for painters and for wall papers.

Tin disulphide can be prepared either in the wet or the dry way; in the wet way, by the action of sulphuretted hydrogen on a solution of tin tetrachloride. The yellow precipitate so obtained has no handsome colour. A far finer product is obtained in the dry way. The process is often regarded as accompanied by particular difficulties, but in reality it is quite simple. It is only necessary in preparing this pigment to take care not to raise the temperature above a certain point, otherwise a great portion of the tin disulphide will be decomposed into sulphur and tin monosulphide. To prevent the temperature from rising too high, an addition of ammonium chloride is made. This salt is volatile at a certain temperature; heat which would otherwise raise the temperature above this point is used in volatilising the ammonium chloride. With a little care it is easy to interrupt the operation before all the ammonium chloride has been driven off. The mosaic gold then obtained has a real golden glitter. If the temperature rises too high, grey tin monosulphide is formed, which naturally considerably diminishes the brilliancy of the product. Ammonium chloride may be replaced by mercury or mercury compounds, which are volatile at a temperature below that at which mosaic gold is decomposed. When mercury compounds are used, both on account of their cost and poisonous nature the heating must be conducted in glass retorts in order to recover the mercury. This operation requires great care if loss due to the breakage of the glass vessels at the high temperature is to be avoided. The process in which metallic mercury is used gives the finest product of all, and is to be recommended when a pigment is to be prepared which shall as nearly as possible approximate to the appearance of gold.

In order to obviate the danger and loss associated with the use of glass vessels, manufacturers who make mosaic gold in large quantity should use an iron vessel. This will last a very long time. Such an apparatus consists of an iron pan with a broad rim, upon which is fastened a head which has the form of a retort neck; to this are connected short, wide tubes leading to a chamber in which substances not condensed in the retort neck may deposit, so that the use of this inexpensive apparatus will not only be without danger, but will be accompanied by the recovery of almost the whole of the volatilised substances. The pan is filled with the materials, the head placed on, the joint tightly luted, and the retort neck connected with the chamber by the wide iron tubes.

There are many formulæ for the preparation of mosaic gold. Some of the most important are given, which in every case will yield a good result:—

Tin filings	40 parts
Sulphur	35 ,,
Ammonium chloride	25 ,,

The tin filings, which must be very fine, are well mixed

in a mortar with the sulphur and ammonium chloride. The heating is gradual at first; when the evolution of vapours has ceased the temperature is very slowly increased to a dark red heat. The mosaic gold is found as a yellow mass at the bottom of the vessel, but partly in crystalline scales on the walls and head of the retort.

Other recipes are as follows :—

Tin dioxide	80 parts
Sulphur	60 ,,
Ammonium chloride	30 ,,
Tin filings	45 ,,
Sulphur	35 ,,
Ammonium chloride	25 ,,

In all these cases the chief endeavour should be not to raise the temperature too high; a dark red heat is sufficient to give a perfectly satisfactory result.

When metallic mercury is used, it is employed in the form of an amalgam with tin, which is then in such a finely divided condition that it readily enters into chemical combination with the sulphur. The amalgam is most simply obtained by heating 1 part of mercury almost to boiling, and stirring 2 parts of tin filings into the hot metal; 18 parts of this amalgam mixed with 7 parts of sulphur and 6 of ammonium chloride are heated together.

The mosaic gold made after any of these methods may be used for gilding gold frames or as a painter's pigment. Much so-called gold paint consists of mosaic gold ground with a thick gum solution.

Chrysean.—Wallach found that when a current of sulphuretted hydrogen was passed through a saturated solution of potassium cyanide a precipitate was formed, which had the formula $C_4H_5N_3S_3$. This substance, chrysean, is similar in appearance to mosaic gold; its technical employment is prevented by its extremely poisonous nature and its high cost as compared with mosaic gold.

CHAPTER XVI.

RED MINERAL PIGMENTS.

VERMILION.

THIS beautiful scarlet red pigment, which has been used for so long a time, consists of mercuric sulphide, HgS. The same compound occurs ready formed in nature as cinnabar; picked pieces of this mineral come into the market under the name of mountain vermilion (*Bergzinnober*). A far larger quantity of vermilion is made artificially.

Mercuric sulphide exists in two forms—as a black non-crystalline powder and in the crystalline form, which is used as a pigment. Each modification may be transformed into the other by suitable treatment, and each may pass into the other spontaneously under certain conditions. In the manufacture of vermilion, the black form of mercuric sulphide plays an important part; it is, therefore, necessary to give an account of the chemical behaviour of the two modifications before proceeding to an account of the method by which vermilion is made.

Black Mercuric Sulphide may be obtained either by the direct union of metallic mercury with sulphur, or by precipitating the solution of a mercuric salt with sulphuretted hydrogen. It is most simply formed by rubbing together equal parts of sulphur and mercury moistened with water, until the mixture is uniformly black. It is, however, difficult in this way to convert all the mercury into sulphide. A better result is obtained when the mixture is moistened by ammonium sulphide instead of water. In this case the time required for the operation is shortened by warming the vessel. If the mortar is placed in hot water it is generally sufficient to grind for about two hours to bring about the combination of the mercury and sulphur.

This compound can also be easily obtained by heating mercury with sulphur. In a vessel, placed under a chimney with a good draught, which is necessary to carry away the poisonous mercury vapours, 6 parts of the metal are heated nearly to boiling and 1 part of sulphur added. Combination takes place at this temperature with a slight explosion, and pure mercuric sulphide results when the heating is continued until the excess of sulphur is driven off.

For the purpose of the manufacture of vermilion, black mercuric sulphide is most simply made by filling a thick-walled vessel with equal weights of mercury and sulphur, moistening the mixture with water and shaking or rotating the vessel for several hours. This can be done either in a rotating cylinder containing iron balls, or the vessel can be fastened to any rotating object—to a water-wheel or to the fly-wheel of a steam-engine. The vessel in which the combination is effected should of course not be quite full. It has been found that a more jerky motion than that of rotation effects the combination of the mercury and sulphur in a shorter time. For example, an opportunity of fastening the vessel to a saw-mill would be of great advantage.

The mercuric sulphide made by the above methods is a velvety black mass, which, even when exactly equivalent weights of sulphur and mercury have been used, is never quite pure. Carbon bisulphide will always extract a certain quantity of uncombined sulphur. The most important property of the black sulphide for the present purpose is that it is changed into the crystalline modification by heating to the temperature at which it volatilises. The sublimed mercuric sulphide has the well-known fiery scarlet colour characteristic of vermilion.

Red Mercuric Sulphide, or vermilion, exhibits, for a sulphur compound, considerable resistance to the action of chemical reagents; dilute mineral acids do not decompose it. Unfortunately, vermilion has another property which makes it quite unsuitable for the artist's use: in the course of time it gradually turns dull and at last is completely discoloured. This alteration of colour can only be ascribed to a return of the red crystalline modification into the black non-crystalline. When a white pigment is tinted by vermilion it should not be a lead pigment, or in a brief time it will turn black. A white pigment such as zinc white, which is not acted upon by sulphur compounds, should be used.

CHAPTER XVII.

THE MANUFACTURE OF VERMILION.

THE red modification of mercuric sulphide can be prepared in the wet or the dry way. The latter was formerly in general use, but at present the wet method is more generally employed, as it more easily and certainly produces a handsome pigment. Each method has its advantages, and both will be described.

(a) DRY METHOD.

The numerous prescriptions which have been given for the manufacture of vermilion by the dry method are all founded on endeavours to convert black mercuric sulphide into the red form. Many of these prescriptions contain directions for the soundness of which no reason can be discovered, and it is not going too far to say that none exists, and that operations described as essential for the success of the process have been inserted merely to give the recipe the appearance of novelty. It cannot be denied that certain manipulations impart a greater brilliancy to the product, although it is impossible to assign a physical or chemical reason; but the manufacturer will quickly be able to differentiate the valuable from the worthless in these processes. Two conditions have the greatest influence on the beauty of the pigment—the temperature at which the black sulphide is sublimed, and the complete freedom of the vermilion from excess of sulphur. Of less importance is the repeated grinding of the vermilion; the oftener it is ground, the smaller the crystals become and the paler the shade.

The operation of grinding the pigment in ordinary mills is known practically as "preparing," the extraction of excess of sulphur by boiling with alkaline liquids as "refining," the product.

The usual process in Holland, especially in the Amsterdam works, is as follows: The black sulphide is made by heating 270 parts of mercury with 37·5 parts of sulphur in copper pans —the fire is so regulated that the temperature is not high enough to ignite the sulphur. If properly prepared the product has now a pure black colour. It is immediately finely ground and preserved in earthenware bottles containing only 0·7 kilogramme each. It is sublimed from hemispherical vessels, provided with iron covers carefully joined to the rim of the sublimation vessel by a suitable fire-resisting cement. Generally 3 such vessels are contained in one furnace. The operation is commenced by heating them until the bottom shows a dark red heat. The temperature should now be increased to such an extent that, when the contents of one of the above-mentioned small bottles are poured into the vessel, a small flame only appears; if on the contrary the contents burn explosively, the temperature is too high, and further quantities of the black sulphide must not be added until the vessels have cooled down to a certain extent. If, on introducing the first quantity of mercuric sulphide, a flame appears unaccompanied by an explosion, the contents of several of the bottles may be introduced; the openings through which this addition is made are immediately closed by a well-fitting iron plate. From time to time this cover is raised and fresh quantities of mercuric sulphide added. The operation lasts about 36 hours, when double the quantity of mercuric sulphide, made from the above mixture of mercury and sulphur, will have been introduced into the 3 vessels. For the complete success of the process, the accurate regulation of the temperature to which the vessels are exposed is particularly important; in practice, the temperature is judged by the height of the flame which issues on removing the iron plates: if this reaches to 1 metre the fire is too fierce, but if the flame is very small the fire must be increased. Towards the end of the sublimation the mass in the vessels is stirred about every 15 minutes. As soon as the sublimation is finished the fire is extinguished. The vessels are broken when completely cold; the vermilion is then found in the upper portion as a sublimate of fibrous character. The vermilion made by this process requires simply wet grinding under ordinary mills and drying to be ready for market.

In the great mercury works at Idria, in Austria, vermilion is also made from the black sulphide. The latter is made by mixing 84 parts of mercury with 16 parts of finely powdered sulphur in rotating vessels driven by water power, the operation lasting about 3 hours. This quantity of sulphur is larger than is required to form mercuric sulphide; experience has shown that the combination takes place more rapidly when more than the equivalent quantity of sulphur is used. Heat is developed by the reaction and the temperature of the mixture rises to over $30°$ C. The black sulphide is then sublimed in cast-iron vessels, which are pear-shaped and built 6 together in a furnace; each holds a charge of 315 kilogrammes of black.

Several periods are distinguished during the heating of the mass in the sublimation vessels. The operation is commenced by heating 2 of the vessels first. As soon as the vapours of sulphur, issuing from the neck of the vessels, take fire with explosion, the fire is made to heat the adjacent vessels. When the contents of all the vessels have exploded, the first part of the process known as the "evaporation" is at an end. Earthenware heads are then placed on the vessels and the fire is increased until the excess of sulphur present begins to distil; its vapours take fire in the air with a slight explosion. When this takes place, earthenware receivers are attached which have only a small opening for the escape of uncondensed vapours. The sulphur condenses in these. When sulphur vapours are no longer given off in quantity, the intermediate period (*stückperiode*) is finished and the real sublimation of the vermilion commences. The fire is now considerably increased and the sublimed vermilion collects in the cooler parts of the apparatus. When the sublimation is completely finished, small blue flames, which quickly vanish, appear at the joints of the apparatus. The furnace is allowed to get quite cold, when the various parts of the apparatus are taken apart, the vermilion deposited in the tubes is carefully removed, so that they may be again used, whilst the receivers and head are broken, so that the vermilion they contain can be collected.

The larger pieces form lump vermilion; the fragments of the receivers are cleaned with a wet brush to collect what adheres. The whole process of sublimation from the introduction of the black to the end lasts about 7 hours.

The sublimed vermilion is ground in mills which differ little from ordinary grinding mills. To prevent the formation of dust, water is added and the stones are surrounded by wooden casings. The red paste from the mills, which is now known as vermilion, is then refined.

The refining consists in extracting the excess of sulphur by means of boiling potash solution; 300 kilogrammes of the ground vermilion are stirred up with water in a tub, the water is drawn off and the wet mass brought into an iron pan, in which it is heated with 22·5 kilogrammes of potash lye for about 10 minutes. According to the composition of the crude vermilion the lye has a strength of from $10°$ to $13°$ B. The smaller the quantity of sulphur, the weaker is the potash solution; it is, or was, made in Idria in a very primitive manner, by extracting wood ashes. For vermilion of a bright red shade potash solution of 10°, for the dark red of 11°, and for "Chinese vermilion" of 13° is used. The excess of sulphur, together with a trace of mercuric sulphide, dissolves in the potash solution; the sulphur chiefly forms potassium pentasulphide. When the boiling is finished, the vermilion is carefully washed and dried in dishes placed in a heated furnace. During drying, the vermilion agglomerates; finally the lumps are broken and sieved.

Chinese Vermilion is in similar case to Indian ink. Both substances are in common use in Europe, they far surpass in quality our own manufacture, and in neither case do we know the exact method by which they are made. Genuine Chinese vermilion so far surpasses European in brightness that it is bought at five or six times the price. It is said on unauthenticated authority to be made by subliming a mixture of 4 parts of quicksilver with 1 part of sulphur in earthenware pots closed by an iron plate, which is kept constantly wet and serves as a receiver, on which the vermilion deposits. The sublimed masses adhering to the lid are sorted, ground and repeatedly washed with water.

According to Callum's description of the manufacture of vermilion at Hong Kong, mercury and sulphur are heated in a large iron pan, with continual stirring at about the melting point of sulphur (111° C.), until the whole has changed to a black mass. After cooling, this is mixed with water and mercury, the mixture thoroughly stirred, dried, placed in a hemispherical dish and covered over with broken porcelain. A similar dish is cemented on the top of the first, and the dishes are heated for 16 hours; the vermilion adhering to the porcelain fragments is removed, wet ground and dried.

A vermilion approaching Chinese in beauty is said to be obtained by mixing ordinary vermilion with 1 per cent. of antimony sulphide and again subliming the mixture. The dark grey sublimate produces a reddish brown powder, which is repeatedly boiled with a solution of "liver of sulphur" (potassium polysulphide), washed with water and digested for a long time with hydrochloric acid. The author has repeatedly made vermilion by this process, but could never obtain a product differing appreciably from the original vermilion.

(b) WET METHODS.

The manufacture of vermilion in the wet way is founded on the conversion of metallic mercury or its compounds into mercuric sulphide by heating with alkaline sulphides, such as ammonium sulphide and liver of sulphur. The product always contains uncombined sulphur, which is eliminated by treatment with alkalis.

Apart from the danger to the health of the workmen caused by the poisonous mercury vapours, which are present in rooms where mercury or its compounds are heated in apparatus even of the best construction, and which is not to be feared in a wet process, the vermilion prepared in the latter manner has a finer and brighter shade than that prepared by a dry process. The greater cost of the wet process is covered by the higher value of the product.

The starting point of the wet method may be either metallic mercury, black mercuric sulphide or another mercury salt. When metallic mercury is used, according to an old process, 100 parts are ground with 23 parts of flowers of sulphur and a little caustic potash solution to a homogeneous mass, which is then heated with a solution of 53 parts of caustic potash in an equal quantity of water, the evaporated water being continually replaced, until the colour changes from brownish red to the fiery red of vermilion. When the colour approaches the desired shade a careful watch must be kept, and immediately the proper shade is obtained the heating must be stopped. If it is continued beyond this point for but a short time, the vermilion at once loses its fire and cannot be again brightened. An interruption in the heating is equally harmful, a dull shade being produced. When the proper shade has appeared, the contents of the vessel are poured into a large quantity of water. The vermilion is then washed with dilute caustic potash and afterwards with water until the alkaline reaction disappears. It is finally dried. The alkaline solutions obtained in this process contain considerable quantities of mercury in solution. They are collected, and when a sufficient quantity has accumulated the mercury is extracted.

Barff recommends the following method: Mercury is rubbed with $\frac{1}{8}$ to $\frac{1}{5}$ of its weight of sulphur until a uniform grey powder results. This is heated in a porcelain dish with caustic potash solution (133 parts of potash in 150 parts of water) at 45° C. until the powder has become bright red. Heating above 45° C. is to be avoided; it would impart a brown tinge to the vermilion.

A convenient process for making vermilion in the wet way is to first produce the black sulphide in the dry way and then treat this with alkalis. Brunner's method is founded on this procedure: 100 parts of mercury and 38 parts of sulphur are used to prepare the black sulphide. This is heated with a solution of 25 parts of caustic potash in 150 parts of water. The operation is best conducted in a vessel placed in a water bath, the temperature of which does not exceed 45° C. After heating 7 to 8 hours, the mass begins to turn red, when the change proceeds more quickly, and the greatest care must be taken not to exceed the point at which the colour has reached its greatest brilliance. As soon as the desired shade has appeared, the temperature of the water bath is lowered, but it is kept warm for several hours. Caustic solutions of different strengths produce different shades. Thus, in order always to obtain the same product from the same quantities, it is necessary to replace the evaporated water at short intervals. In this process some quantity of mercury remains uncombined; it is separated from the vermilion by a process of levigation in the washing

Firmenich's Method.—Many wet methods for the manufacture of vermilion differ but little from the preceding. In Firmenich's method the production of the black sulphide is united with the formation of vermilion in one operation.

A solution of potassium sulphide of a certain strength is made by igniting 20 parts of potassium sulphate with 6 parts of coal, boiling the mixture with 1·5 time the quantity of rain water, cooling the solution, separating from the potassium sulphate which crystallises out, and boiling the solution with sulphur so long as it is dissolved. A mixture of 4·5 parts of this solution, 100 parts of mercury, and 2 parts of sulphur is placed in flasks, which are subjected for several hours to a shaking or rocking motion. The liquid becomes greenish, and its temperature rises in consequence of the combination of mercury with a portion of the sulphur of the potassium sulphide. The latter immediately dissolves the free sulphur present, and again gives it up to the mercury, which in the course of a few hours is completely converted into the black sulphide. The shaking is discontinued when the contents of the flasks have acquired a deep brownish red colour. The flasks are then placed in a room heated to 45° C., and their contents repeatedly shaken up. At this temperature the transformation of the black into the red mercuric sulphide is completed. The deposit in the flasks acquires more and more a scarlet shade. As soon as the colour is found to have reached its greatest intensity the liquid is carefully poured off, and the vermilion treated with caustic soda solution in order to dissolve free sulphur; it is then very carefully washed.

In this process, the temperature at which the operation commences is important. The cooler the mixture which is placed in the flasks before shaking, the paler will be the the colour of the vermilion obtained. It is to be supposed that the reaction, which is but slow in the cold, forms in consequence of its slowness a black sulphide of such character that in its subsequent transformation into the red crystalline sulphide very small crystals are produced.

The Gautier-Bouchart method for the preparation of vermilion from mercury and ammonium sulphide is applied on the large scale as follows: 1,000 parts of mercury are shaken for 7 hours with 200 parts of flowers of sulphur and 400 parts of ammonium sulphide saturated with sulphur; the dark coloured mixture is exposed for several days to a temperature of 60° C., when the colour changes to red. In addition to the usual washings, the mass is further treated with nitric acid. The purpose of this operation is to oxidise all the free sulphur to sulphuric acid.

For a colour works in which, in addition to vermilion, are made other pigments, especially such as are sensitive to the action of sulphuretted hydrogen, this process, which produces vermilion of a good shade, although not particularly stable, is not to be recommended on account of the evolution of sulphuretted hydrogen, which cannot be completely prevented even by the greatest care.

Liebig's Process.—The starting point in the manufacture of vermilion by this process is the compound known as white precipitate (see below). It is only necessary to heat white precipitate with a solution of ammonium sulphide saturated with sulphur, at 40° to 50° C., for a long time. The operation may be conducted in well-closed flasks in a place the temperature of which is, as nearly as possible, 45° to 50° C., such as the flues leading from a furnace in constant use. The change of colour is gradual; this is an advantage, since it is easier to obtain the correct shade. By treating the product with weak caustic potash solution it may be made still brighter. After washing and drying at a gentle heat the pigment is finished.

This method is especially adapted for manufacturers not exclusively engaged in making vermilion. No special arrangement of apparatus is necessary; the apparatus for making ammonium sulphide and a number of glass flasks are the only essentials. There is another considerable advantage, that the product cannot be completely spoiled. The process of formation of the vermilion is tolerably slow; a careful observation of the progress of the earlier operations is sufficient to determine the time required. In order to be able to do this with certainty, it is necessary to work always under exactly the same conditions; the same quantity of white precipitate must always be used and the solution of ammonium sulphide must always have the same strength.

Mercuric Ammonium Chloride, Infusible White Precipitate, $HgClNH_2$.—This compound, which is required in the last-mentioned process for the manufacture of vermilion, is most cheaply made in the following way: to a solution of 1 part of common salt in 32 parts of water, 2 parts of dry mercuric sulphate are added in small quantities, whilst thoroughly stirring. It is absolutely necessary to work in this way, because mercuric sulphate is decomposed by water into free sulphuric acid and a basic salt, which is much more slowly converted into white precipitate than the neutral salt. The liquid now contains mercuric chloride; on the addition of ammonia to alkaline reaction it gives a heavy white precipitate. The liquid is poured off and the precipitate washed with water containing a little ammonia, until the washings give only a slight turbidity with barium chloride. This process may also be commenced with mercuric chloride, but the above method is cheaper. White precipitate should volatilise without fusing when heated on platinum foil and should keep its white colour when treated with ammonia.

Electrolytic Process.—In a wooden vessel 1 metre in diameter and 2 metres deep are placed, at the sides, dishes 15 centimetres wide, containing a layer of mercury 1 centimetre deep. These dishes are connected with the positive pole of a dynamo. The negative pole is connected to an iron plate, electrolytically coated by copper, placed at the bottom of the vessel, which is filled with a solution containing 8 per cent. of ammonium nitrate and 8 per cent. of sodium nitrate. A regular current of sulphur dioxide is introduced through a perforated coil. The excess of gas escapes by a tube in the cover. When the current is passed a precipitate of red mercuric sulphide is at once formed. Attempts have been made to dispense with the current of sulphur dioxide The bath then contains 100 litres of water and 8 kilogrammes each of ammonium nitrate, sodium nitrate, sodium sulphide and sulphur. Under these conditions it is only necessary to add sulphur and mercury in order to obtain, at the end of the operation, vermilion which is in no way inferior to that made by the first process.

Vermilion is frequently grossly adulterated by cheaper pigments. Substances are often found under the name of vermilion which contain no trace of mercury, but consist of bright orange lead mixed with a few per cent. of ferric oxide, and having a deceptive similarity in colour to the best vermilion.

Mercuric Iodide.—When corrosive sublimate solution is precipitated with exactly the necessary quantity of potassium iodide, mercuric iodide is obtained as a scarlet precipitate which surpasses in beauty even the best samples of vermilion. Unfortunately, this substance cannot be used as an artists' colour. Exposure to light soon turns it brown and finally black. It appears to be unaltered in the dark; the author possesses a sample which has been so kept for 30 years without losing its shade in any way.

CHAPTER XVIII.

ANTIMONY VERMILION.

ANTIMONY VERMILION is a red pigment which will bear comparison in fineness of shade with mercury vermilion, over which it has the advantage of cheapness. In composition it is antimony trisulphide, Sb_2S_3. This compound is obtained by precipitating a solution of antimony trichloride with sulphuretted hydrogen. However, the precipitate, which is a very fine red whilst wet, loses its colour in drying, and the product is almost worthless as a pigment.

In another way it can be obtained in such a condition that it loses nothing of its beauty in drying, but retains its brilliance. Böttger gives the following process: a solution of antimony trichloride is mixed with a solution of sodium hyposulphite (thiosulphate) and the liquid heated so long as a precipitate forms, which is then washed on a filter with water containing acetic acid. If pure water were used for washing, the antimony chloride still present would be decomposed, forming the white oxychloride, which would detract from the shade of the antimony vermilion. In this process particular regard is to be paid to the use of exact quantities of materials. The finest product is obtained when 2 parts of a solution of antimony trichloride, which has exactly the specific gravity 1·35, are mixed with a solution of 3 parts of sodium hyposulphite in 6 parts of water.

According to R. Wagner, antimony vermilion is obtained by dissolving 4 parts of tartar emetic and 3 parts of tartaric acid in 18 parts of water, heating to 60° C., mixing with a solution of sodium hyposulphite (thiosulphate) and heating to 90° C. The precipitate is then carefully washed and dried.

Pure antimony vermilion closely approaches, as we have said, ordinary vermilion in shade, and for a sulphur compound shows a remarkable resistance towards chemical reagents. By dilute acids, ammonia and alkaline carbonates, it is attacked only on long-continued contact, but it is easily decomposed by very dilute hydrochloric acid and by caustic alkalis. A mixture with white lead keeps for a long time, but there can be no question of the permanence of such a mixture, in consequence of the oft-repeated properties of lead pigments. Antimony vermilion is well adapted for oil painting. When ground with oil it exhibits a red of a brilliance in no way inferior to that of genuine vermilion. It may also be used as a water colour, but is not adapted for fresco work, since it is quickly decomposed by lime.

In spite of these favourable properties antimony vermilion has so far been little used. Considering the small cost of its preparation, especially when calcium hyposulphite, which gives an equally good result, is used in place of the sodium salt, its use is to be recommended in the place of mercury vermilion. It appears as if the high price demanded for this substance by several manufacturers has prevented its general employment.

Appendix. Antimony Blue.—This fine blue pigment can be prepared by the addition of a dilute solution of potassium ferrocyanide to a clear solution of antimony in *aqua regia*. According to Krauss, it contains no antimony as colouring principle, but is a Prussian blue obtained from the ferrocyanide, which is decomposed by the strong acid, with evolution of hydrocyanic acid.

CHAPTER XIX.

FERRIC OXIDE PIGMENTS.

THE pigments composed of ferric oxide are used in enormous quantity. They are distinguished by a high degree of permanence. Large deposits of ferric oxide occur in nature, and in places where it is found in considerable quantity iron is manufactured from it on the spot. Several varieties of natural ferric oxide are distinguished: specular iron ore forms crystalline masses of brilliant lustre; another variety in small crystals is called iron glance; micaceous iron ore consists of shining scales; red hæmatite has the appearance of bundles of fibres; an earthy variety of hæmatite is also common.

The compound of ferric oxide with water, ferric hydroxide, is still more abundant than hæmatite; brown hæmatite, limonite and other iron ores consist essentially of this compound. The pigment known as ochre is also ferric hydrate.

Very pure red hæmatite has so fine a red colour that it may often be used as a pigment after grinding or levigating. The famous red of Pompeii, which has been found on the ruined walls of the town, still shows, after eighteen centuries' exposure to damp, the brightest shade, certainly a striking proof of the extraordinary permanence of this pigment. Considering its great permanence, its easy preparation, and its low price, it is surprising that ferric oxide is not more extensively used by artists than it is at present. It is used,

FERRIC OXIDE PIGMENTS. 181

however, extensively in colouring earthenware, for which purpose its stability at high temperatures makes it suitable. Principally on account of its cheapness it is largely used in ordinary painting, but for artistic purposes it is not used to the extent it deserves.

Every colour maker well knows that artists justly complain that the pigments offered to them have generally but a small degree of permanence. They are surprised that the paintings of the old masters show now, after the lapse of centuries, their colours unaltered, whilst the pigments manufactured at the present day, instead of corresponding to the high standard of chemical knowledge, are often discoloured within a few months after use. But it was just the restricted knowledge of chemistry which the ancients possessed which compelled them to make extensive use of the permanent earth pigments, to which class of permanent colours ferric oxide belongs. The advances of science have succeeded in preparing ferric oxide, not only as a red pigment, but in the different shades of red, from yellow to brown and deep violet, all consisting entirely of pure ferric oxide. Ferric oxide has the property of altering its molecular condition on protracted heating; this change is accompanied by an alteration in colour. If ferric oxide is heated for a very long time at the highest temperatures its colour changes at last to black.

Ferric oxide can be prepared by different methods for artists' purposes. The process by which it is made is not unimportant. Either ferrous or ferric salts can be employed as the raw material. With the latter, pure ferric oxide is at once formed, or hydroxide, which is converted into oxide by heating. The ferrous salts are generally cheaper than ferric salts; they are therefore commonly used for the preparation of ferric oxide, as well as of the other iron pigments. Even in combination with the strongest acids, ferrous oxide has but little stability; when separated from its salts as ferrous hydroxide, the greatest precautions must be taken to obtain it pure; in contact with the air it at once takes up oxygen and changes to ferric hydroxide. Ferrous carbonate shows this same degree of instability; the natural substance, occurring in large crystals as spathic ironstone, is no exception: on exposure to the air it is gradually changed to ferric oxide.

There is another reason for the advisability of using ferrous salts to prepare ferric oxide. When ferric hydroxide is made by precipitating the solution of a ferric salt with ammonia or caustic potash, the least excess of the precipitant unites with the hydroxide to form a compound which is only decomposed by long washing with water. The precipitate is, however, gelatinous, and consequently very difficult to wash.

In order to prepare ferric oxide suitable for an artists' pigment the following process may be used: 17 parts of soda are dissolved in 68 parts of water; the solution is boiled in an iron pan, and 10 parts of crystallised ferrous sulphate are added in small quantities with continual stirring. The boiling and stirring are continued until the green vitriol has completely dissolved, when the greenish white precipitate is allowed to settle, washed several times with water, and then exposed to the air in thin layers. The precipitate, which begins to turn yellow during washing, becomes in a short time ochre yellow in the air, being changed into ferric hydroxide. After drying and calcining, a fine red powder of pure ferric oxide is formed. The shade depends on the temperature at which the substance is calcined: the higher the temperature and the longer the heating is continued, the darker is the product.

Vogel's Iron Red.—This preparation, which is particularly brilliant and therefore highly suitable for an artists' colour, is made by adding a saturated solution of oxalic acid to a boiling solution of green vitriol. The greenish yellow precipitate of ferrous oxalate is collected on a filter and well washed with water. After drying, the precipitate is heated in a shallow iron dish to a temperature of 200° C., at which the ferrous oxalate decomposes and is converted into a soft fiery red powder consisting of pure ferric oxide. By igniting this powder in covered crucibles the different shades of ferric oxide can be obtained.

Macay's English Red.—Seven hundred and four parts of ferrous sulphate, 1,000 parts of copper chloride, and 1,678 parts of common salt are dissolved, the solution boiled and the precipitate ignited.

In the manufacture of fuming or Nordhausen sulphuric acid, ferric oxide is obtained as a residue. It is then known under the names of English red, *caput mortuum*, colcothar, rouge and Indian red as a very cheap pigment. It is also used as a polishing material. Fuming sulphuric acid is made by heating green vitriol at a white heat in retorts placed in furnaces; sulphur dioxide and trioxide are evolved, whilst in the retorts there remains a residue of almost pure ferric oxide, containing small quantities of basic ferric sulphate, which can only be decomposed by long continued violent heating. The vapours of sulphur trioxide are caught in receivers containing oil of vitriol, in which they dissolve and produce fuming sulphuric acid.

The residue in the retorts, which has only a low commercial value, can be converted into a good pigment without the expenditure of much money or labour. It is ground in mills as finely as possible, and, if necessary, afterwards levigated. The fine powder is mixed with varying quantities of common salt, the object of which is to prevent the temperature from rising too high in the calcining process. Common salt is volatile at a temperature approaching a strong red heat; when the temperature has once risen so far, a further rise is prevented by the heat taken up in volatilising the salt. In order to make ferric oxide of a yellow tinge, 2 per cent. of salt are added, and the mixture heated with a moderate fire for 1 hour. To obtain the deeper shades, the addition of common salt is increased even to 6 per cent. For a brownish red oxide, 4 per cent. of salt are added and the mixture heated for 4 hours; for a dark violet oxide, 6 per cent. of salt are used, and the mixture is heated for 6 hours with the fiercest fire.

It has been observed that the shade of the product is finer the more completely air is excluded from the glowing mass and the more quickly the product, after sufficient heating, is cooled down to the ordinary temperature.

Fire-clay tubes are used for heating the ferric oxide, which are similar to gas retorts, and are built one above the other in furnaces. The number in one furnace may reach 60. Each retort is closed by a well-fitting lid, which is luted with clay after the retorts are filled, a small opening being left through which the heated air may escape. If the retorts were closed completely air-tight they would burst on heating.

Ferric oxide pigments are made in very large quantity by several works, which bring up to 20 different shades, varying from reddish yellow to dark violet, into the market.

On account of their great resistance to the action of the atmosphere and of chemical agents, the ferric oxide pigments are particularly suitable for painting iron and other metals which are exposed to air or water. They are also suited to fresco work.

Lower qualities of iron reds are made by calcining ochres, large deposits of which occur in nature, and also from the residues of basic ferric sulphate obtained in the alum manufacture. These varieties are at the best usable for ordinary painting, but never for artists' pigments.

According to the method of Steinau, iron pigments can

be made from wrought-iron scraps, turnings, etc., by causing them to rust through alternate contact with air and water. The resulting ferric hydrate is either at once used as a pigment or converted into other shades by calcining. By heating in the air a red pigment is formed ; by heating with coal in the absence of air a black ; and from mixtures of red and black different shades of brown. If the iron turnings can be obtained cheap, this process should be well adapted to the manufacture of fine iron pigments.

Indian Red consists of ferric oxide. It was originally obtained from a very pure hæmatite, occurring in India, by grinding and levigating. It can, however, also be artificially made, in shades varying between bright red and dark brownish red, by heating pure ferric oxide. This valuable pigment is extremely brilliant and durable.

The darkest brownish red shades can be obtained from Indian red by mixing it with varying quantities of litharge, and heating the mixture very strongly in a covered crucible.

CHAPTER XX.

OTHER RED MINERAL PIGMENTS.

Chrome Red or Chrome Vermilion.—Under the most varied names, chrome orange, Persian red, Derby red, Chinese red, Indian red, chrome vermilion, etc., there occur in commerce numerous pigments, orange to dark red in colour, which can all be made from neutral lead chromate. When this salt, *i.e.*, ordinary chrome yellow, is treated with smaller or larger quantities of a strong base, such as caustic potash or caustic soda, "basic" lead chromates are formed, which exhibit a more intense red shade the more lead they contain. These compounds can be made either by adding caustic potash to the solution of potassium chromate used for making the chrome, or the precipitated chrome yellow may be treated with caustic potash. The latter process readily yields good results, so that it is generally followed.

Recent researches have shown that differences in the shade of chrome red are due to the varying size of the single microscopic crystals of which it consists. When chrome reds of the most different shades are ground the powder has almost the same colour. The larger are the crystals the deeper is the shade of the red, so that the art of making deep chrome red lies in working so as to produce large crystals.

As was said in describing the preparation of chrome yellow, the lead liquors used in its manufacture contain varying quantities of acetic acid, which affect the shade of the product. The quantities of caustic potash necessary to turn a certain quantity of chrome yellow into chrome red of a certain shade can, therefore, only be estimated from time to time in an empirical manner. This estimation may be performed by the method of Habich with great certainty. The precipitate of chrome yellow is made in the ordinary way, and well washed 6 to 8 times. Equal portions of the paste are brought into vessels of equal height and diameter. To the first of these portions caustic soda solution of a certain strength is added; to the second rather stronger caustic soda solution; in each succeeding vessel the strength of the solution is increased to a known extent. After the addition of caustic soda the vessels are energetically shaken. They are then set aside for the precipitate to settle in a place completely free from vibration. In one of the vessels, an orange or red shade will be seen very similar to that it is desired to make. Now, it is known how much caustic soda solution of a certain strength has been added to this quantity of colour, so that the quantity necessary for the whole of the chrome yellow precipitate may be found by a simple calculation. In working on the large scale, the process is exactly the same as in the preliminary test; the calculated amount of caustic soda solution is added to the precipitate in the tub, it is well stirred, and left at rest for some hours. When the proper shade appears, the liquid above the precipitate, which contains sodium chromate in proportion to the strength of the caustic soda solution used, can be drawn off and used to precipitate fresh lead acetate solution. The precipitate is washed first in the tub and then on the strainers.

Since, as has been said, the depth of colour depends upon the size of the single crystals, care should be taken in washing the precipitate not to agitate it too vigorously, or the pigment will lose in depth by the breakage of a large number of crystals.

Special formulæ for the manufacture of chrome red are abundant, but if the above process be followed they are superfluous, and it is only for the sake of completeness that one or two are given. Deep chrome red is made by the action of 25 parts of caustic soda on 100 parts of pure chrome yellow.

A noteworthy process is due to Liebig. Saltpetre is melted in a crucible, and heated to a temperature below that at which oxygen is evolved. Dry chrome yellow is introduced so long as effervescence occurs; the fused mass appears, whilst liquid, deep black. As soon as it is in quiet fusion it is poured on a cold plate. In consequence of the rapid cooling it becomes brittle and can be easily powdered. The hot mass is broken up and boiled with water, potassium chromate going into solution. The chrome red made by this process is little inferior in brilliance to vermilion.

According to Prinveault a handsome chrome red is obtained by treating 25 grammes of neutral lead carbonate with 10 grammes of yellow potassium chromate dissolved in 5 litres of water, boiling for half an hour, washing the violet precipitate, and finally treating with 1 gramme of sulphuric acid diluted with 100 grammes of water.

Cobalt Red consists of cobalt phosphate. It was first recommended by Salvetat. It is obtained by precipitating a solution of a cobalt salt by sodium phosphate. After drying at the ordinary temperature the precipitate has a beautiful rose red colour. By careful heating, the shade becomes more violet according to the temperature used.

Cobalt-Magnesia Red. — When magnesia is moistened with the solution of a cobalt salt and strongly heated, a rose red mass is obtained which can be used as an extremely durable artists' colour. It probably consists of a compound of cobalt oxide with magnesia. The method of preparation is as follows: Magnesium carbonate is mixed to a thin paste with a dilute solution of cobalt nitrate; this is heated and stirred until quite dry, when the residue is strongly heated in a covered crucible. In the manufacture of this and other cobalt pigments it is of particular importance to prevent the entry of the fire gases into the crucible. Unless this is done a product of good colour is not obtained.

Cobalt Arsenate. — By precipitating a cobalt salt with sodium arsenate, a violet-red precipitate is obtained which leaves a fine durable red on heating. The mineral erythrite has a similar composition and colour. This pigment has a bright shade, is permanent, and although it contains arsenic is not particularly poisonous, since after fusion it is little soluble.

Chromium Stannate. — This artists' pigment, which may be used in oil and porcelain painting, is made in the following manner (Gentele): 1 kilogramme of tin is converted by strong nitric acid into metastannic acid, 50 grammes of potassium chromate are dissolved in 1 litre of water, the solution mixed with 2 kilogrammes of chalk and 1 kilogramme of powdered quartz, and finally with the metastannic acid. The mixture is dried and strongly heated. The temperature must be raised to a white heat, at which the mass sinters and acquires a dark rose-red colour. It is then completely extracted with boiling water, when a fine powder is left.

This pigment, which is known in the market as "pink colour," may be used with advantage in oil painting in the place of rose madder lake, over which it has the advantage in durability.

Silver Chromate. — A solution of silver nitrate gives, with potassium chromate, a deep red precipitate of silver chromate, which is occasionally used as an artists' pigment under the name of "purple red". Not only is the price of this pigment high, but it has little durability, since, like all silver compounds, it is acted upon by sulphuretted hydrogen and turned dark, owing to the formation of black silver sulphide.

CHAPTER XXI.

PURPLE OF CASSIUS.

PURPLE OF CASSIUS, or gold purple, is not a pigment for painters' use; it is employed for colouring glass, for porcelain painting and for coloured glazes. For these last purposes we know nothing which can completely replace purple of Cassius, so that its preparation will be described here. Its costliness prevents its general use.

As regards the chemical constitution of purple of Cassius very diverse views are held. Chemists have not yet succeeded in explaining it; some regard it as aurous stannate, others as stannic acid, in which a particular form of gold of a red colour is contained in an extremely finely divided condition. There are reasonable grounds for both opinions. Theoretical considerations are of little interest to the manufacturer, who is chiefly interested in the method by which he can prepare a product which satisfies the demands of the consumer.

Purple of Cassius has been known for nearly two hundred years, hence there are many formulæ for its preparation. Before these are enumerated it is necessary to give an account of the conditions under which this substance is formed. In order to obtain gold purple, a solution is required which contains both stannous and stannic chlorides; if this solution is mixed with a very dilute solution of gold chloride a precipitate is obtained which is generally brownish red, and only acquires its fine red shade on igniting. The colour shown by the purple after precipitation is no guide to the shade of the finished product; a purple which shows a very fine colour whilst wet often produces a pigment of much less beauty than an actually ugly precipitate. These differences are to be ascribed to the varying molecular condition of the purple, for the different varieties show very small differences in chemical composition.

According to the directions of Fuchs, a fine purple is obtained by mixing a solution of stannous chloride with sufficient ferric chloride solution to give the mixture a green colour. Part of the ferric salt is then reduced to ferrous salt, whilst the stannous chloride is partially oxidised. At the same time a solution of gold chloride is made; this must be quite free from nitric acid and contain 1 part of the salt in 400 to 500 parts of water. The tin solution is added drop by drop to the gold solution with constant stirring; the mixture becomes turbid, but the precipitate requires a long time to settle on account of its fine state of division. The depth of colour shown by the purple varies according to the strength of the solutions.

M. Müller states that the following process gives the best results: The quantity of stannic chloride equivalent to 9 grammes of stannic acid is dissolved in 200 cubic centimetres of water; potassium carbonate is added to alkaline reaction, then 1 gramme of gold in the form of chloride. Grape sugar is added to the mixture, which is made up to 300 cubic centimetres by water, and warmed until the brightest shade is reached. When a very gelatinous mass is formed after the addition of the potassium carbonate it is heated for a short time before adding the gold solution and grape sugar.

Wächter gives directions for obtaining pale and deep shades. To obtain the pale purple 5 grammes of tin are dissolved in *aqua regia*, the solution evaporated to dryness on the water bath, the solid residue mixed with a solution of 2 grammes of stannous chloride and dissolved in 10 litres of water. To this liquid is added a solution of 0·5 gramme of gold in *aqua regia*, and immediately 50 grammes of ammonia to neutralise free acid. The precipitate separates spontaneously from the dark red liquid; its formation may be hastened by adding sulphuric acid. The washed precipitate is mixed whilst moist by means of a silver spatula with 20 grammes of lead flux, 2 grammes of red lead, 1 gramme of quartz sand, and 1 gramme of calcined borax. The mixture is then dried; it produces a purple red colour. If 3 grammes of silver carbonate are added, a clean, pale purple pigment results.

The deep purple is made by mixing a solution of 0·5 gramme of gold in 10 litres of water with 7·5 grammes of stannous chloride solution (specific gravity, 1·7) whilst stirring, and adding a few drops of sulphuric acid. The washed precipitate is mixed with 10 grammes of lead flux and 0·5 gramme of silver carbonate.

A rose-red shade of gold purple is obtained by dissolving 1 gramme of gold and mixing the solution simultaneously with solutions of 50 grammes of alum in 20 litres of water, and of 1·5 gramme of stannous chloride (specific gravity, 1·7). Ammonia is now added until all the alumina is precipitated. In order to prepare from the dried precipitate the mixture which will produce the colour on fusion, it is mixed with 50 grammes of lead flux and 2·5 grammes of silver carbonate.

Magnesia Gold Purple. — According to M. Müller, the colour of gold purple is produced by covering the particles of a very finely divided white substance with metallic gold. Thus, when calcined magnesia is stirred up with water, gold chloride solution added, and the mixture warmed to 100° C., the gold is precipitated upon the magnesia; the wet, yellow powder acquires a reddish shade on drying, which at a red heat becomes so beautiful a carmine-red that it surpasses purple of Cassius in fineness of shade. A purple which contains 20 grammes of gold as chloride to 84 grammes of magnesia has the most pure carmine tint; the shade varies with the proportion of gold to magnesia. The different shades of purple contain the following percentages of gold :—

Percentage of Gold.	Shade of the Purple.
33·5	Brownish red (excess of gold).
25·0	Deep carmine-red.
20·0	Medium carmine-red.
10·0	Pale carmine.
5·0	A good rose-red.
3·0	Rose-red.
1·0	Pale rose.
0·2	Delicate rose.
0·1	Appreciable red tint.

Alumina Gold Purple is obtained, according to Müller, by adding gold chloride and excess of potassium carbonate to alum solution and boiling. The purple obtained in this way is as deep in colour with 10 per cent. of gold as a magnesia purple with 20 per cent. of gold, but has a more bluish violet shade. If, in the preparation of purple, the alumina is precipitated by potassium carbonate and the gold chloride reduced by grape sugar, a different shade of purple is obtained to that in the precipitation of which ammonia is used. The alumina gold purple is specially adapted for the production of porcelain enamels.

CHAPTER XXII.

BLUE MINERAL PIGMENTS—IRON-CYANOGEN PIGMENTS.

The pigments occurring in commerce under the names of Prussian, Chinese, Berlin or Paris blue consist, when pure, of ferric ferrocyanide, $Fe_4[Fe(CN)_6]_3$. This compound is obtained as a deep blue precipitate when yellow prussiate (potassium ferrocyanide) is mixed with the solution of a ferric salt.

The precipitate shows certain differences in composition according to the exact method by which it is obtained. A different result is obtained when the solution of the ferric salt is poured into that of a ferrocyanide, to that produced by the reverse procedure, the precipitate produced in the one case has properties which it does not possess in the other. Similar differences often occur in precipitating pigments.

If the solution of the ferrocyanide is poured into the iron solution, the latter being in excess, an insoluble precipitate is formed; but if the iron solution is poured into the ferrocyanide solution of which an excess is present, the blue precipitate is formed, but it is soluble in pure water, though not in water containing salts. Thus when this precipitate is separated from the liquid and washed, the washings are at first quite colourless and remain so whilst salts are present, but when these have been completely washed away, a fine blue solution of the ferric ferrocyanide is formed, from which the addition of a salt solution again separates the dissolved Prussian blue.

Before proceeding to the manufacture of Prussian blue on the large scale, it is necessary to consider the behaviour of ferrous salts towards yellow prussiate, since these are commonly used instead of ferric salts. When a solution of potassium ferrocyanide is mixed with a solution of a ferrous salt, there is formed a pure white precipitate of potassium ferrous ferrocyanide, $K_2FeFe(CN)_6$. In order to obtain a completely white precipitate, it is necessary that the ferrous salt should be completely free from ferric salts and that the solutions should contain no dissolved oxygen; the solutions would therefore have to be boiled before mixing. If one of the solutions contains the smallest quantity of oxygen a bluish precipitate is formed. When the white precipitate of potassium ferrous ferrocyanide is exposed to the air it immediately acquires at the surface a dark blue colour, and is completely changed to Prussian blue by sufficiently long contact with air.

In accordance with this behaviour of ferrous and ferric salts different methods may be used to obtain Prussian blue. Either the process is commenced with a ferric salt from which a precipitate of Prussian blue is at once obtained, or the solution of a ferrous salt is converted into ferric salt by a powerful oxidising agent, such as chlorine or nitric acid, or a solution of a ferrous salt is precipitated with yellow prussiate and the white or bluish precipitate, which consists chiefly of potassium ferrous ferrocyanide, converted into ferric ferrocyanide or Prussian blue by treatment with nitric acid.

Prussian blue is also known, as we have said, as Chinese blue, Berlin blue and Paris blue; Brunswick blue or mineral blue is another form. Generally the product under the name of Chinese or Paris blue, which has an intense dark blue colour, is quite pure, whilst under the names of Prussian blue and Berlin blue mixtures with starch or alumina are included, which consequently have a more or less pale colour. The pigments known as mineral or Brunswick blue are the least valuable, they have generally a less deep colour and are quite without the metallic lustre which distinguishes Chinese blue and the better varieties of Prussian blue. Mineral blue often contains very large quantities of alumina, chalk, or even barytes; the addition of the last is to be regarded as irrational, since it makes the pigment conspicuously heavy.

Chinese Blue.—Pure Chinese blue appears in the form of deep blue masses, characterised by a peculiar metallic lustre, which is especially marked when the surface of a lump is rubbed with the finger nail. This metallic lustre is accompanied by a copper-red shimmer which is similar to that exhibited by fine indigo. In large pieces pure Chinese blue appears very dark by artificial light; it possesses enormous strength of colour. The following test serves for the recognition of the complete purity of Chinese blue: A small quantity is powdered and rubbed in a thin layer on white paper; if the metallic lustre shows in undiminished strength the blue may be regarded as quite pure, for Chinese blue quickly loses this property by additions of other substances.

From what has been said above concerning the different behaviour of ferrous and ferric salts, the simplest method of making Chinese blue would be to precipitate the solution of a ferric salt with yellow prussiate, keeping the ferric salt in excess, to wash and dry the precipitate. Ferric salts are, however, dearer than ferrous salts, so that the latter are generally used.

In making Chinese blue from a ferrous salt ferrous sulphate or green vitriol is generally taken. This salt is dissolved in water; it is advisable to add a small quantity of sulphuric acid, since the water will contain carbonates, which produce ferrous carbonate. Air oxidises the latter to ferric hydroxide, which is brown, and if present even in small quantity will spoil the shade of the blue. This addition of sulphuric acid is of special importance when the white precipitate produced by the prussiate solution is to be oxidised by the air.

A solution of yellow prussiate is added to the iron solution in such quantity that a very small excess of iron salt is left ; there results a white or rather, since the ferrous sulphate always contains small quantities of ferric oxide, a pale blue precipitate. This is allowed to settle, the liquid drawn off as completely as possible from the precipitate, and nitric acid then added, or a solution of bleaching powder, followed by the amount of sulphuric acid required to decompose it. The nitric acid or the chlorine liberated from the bleaching powder speedily effects the oxidation, changing the pale blue colour to the deep blue of Chinese blue. The precipitate should be left for several days in contact with the liquid in order to complete the oxidation ; it is then well washed and dried. Pieces of definite shape and size bearing the trade mark of the firm are often pressed from the blue when it has acquired a stiff consistency.

There are many formulæ for the preparation of Chinese blue, differing in regard to the quantities of green vitriol, yellow prussiate, and nitric acid or bleaching powder to be used. It is, however, clear that only *one* formula can be correct, that in which the quantities of materials are equivalent, since the reactions take place between equivalent quantities. In practice the equivalent quantities would not be weighed to the tenth of a gramme—there would be no object in such a proceeding, since in works chemically pure substances are not used ; but manufacturers should work according to equivalent quantities in order to use up their materials as completely as possible. No increase in labour is involved, the same labour weighs one or another quantity.

The proportion of materials in which the least loss is

involved is here given. Nine kilogrammes of green vitriol are dissolved in 100 litres of water, 15 kilogrammes of strong sulphuric acid are added, and then a solution of 15 kilogrammes of yellow prussiate in 100 litres of water. The solutions are kept in constant motion during the mixing. As soon as the decomposition is finished, without waiting for the precipitate to settle, steam is led in and 20 kilogrammes of nitric acid of 1·3298 specific gravity added in small portions. The heating by direct steam is continued until red vapours are no longer evolved from the liquid, which is a sign that the oxidation is finished.

Although the use of direct steam is very convenient for heating, since wooden vessels can be used and the heating quickly accomplished, it may be dispensed with, and the solutions heated in pans or even not at all. In this case the oxidation lasts considerably longer than when the liquid is heated. In the process recommended by Gentele 109 parts of yellow prussiate are used to 20 parts of green vitriol, each dissolved in much water. The precipitate is heated for a short time with 51 parts of nitric acid of 1·2285 specific gravity and 16 parts of sulphuric acid. The mixture is allowed to stand for several days to complete the oxidation before the precipitate is separated from the liquid.

In Hochstätter's method, solutions of 6 parts of yellow prussiate in 15 parts of water and of 6 parts of green vitriol in 5 parts of water are mixed. To the precipitate 24 parts of strong hydrochloric acid and 1 part of sulphuric acid are added, and then a solution of bleaching powder in 80 parts of water until the liquid smells distinctly of chlorine. Apart from the fact that this method does not employ equivalent quantities of iron salt and yellow prussiate, it is also to be regarded as unsuitable, since the addition of sulphuric acid produces calcium sulphate, which, being soluble with great difficulty, will be precipitated along with the blue, making treated with a mixture of quicklime and potassium sulphate. Potassium ferrocyanide is formed; it is obtained in crystals by the evaporation of the solution and can be used in the preparation of Prussian blue.

The "Laming's mass," now generally used in gas works, contains, after use, a considerable quantity of cyanogen compounds, and it would only require a short trial to decide whether it was more advantageous to make this mixture by exposure to air again suitable for the purification of coal gas, or to use it to obtain yellow prussiate.

Turnbull's Blue.—When a solution of a ferrous salt is precipitated by potassium ferrocyanide, a fine blue precipitate is formed, identical in physical properties with Chinese blue, but differing in chemical constitution. It is ferrous ferrocyanide, $Fe_3[Fe(CN)_6]_2$. The manufacture of Turnbull's blue is expensive; the pigment has no specially advantageous properties and is therefore seldom made.

In preparing this pigment, a solution of yellow prussiate is transformed into red prussiate by passing chlorine through it so long as it is absorbed by the liquid; the mother liquors from which red prussiate has been crystallised may be used with advantage. When to the solution of red prussiate green vitriol solution is added so that the former remains in excess, a soluble Turnbull's blue is obtained, but when excess of green vitriol is used the insoluble form is produced.

Chinese blue and Turnbull's blue are used both as oil and water colours, and these handsome but not particularly durable pigments are also employed in colouring wall papers.

Antwerp Blue is a mixture of Prussian blue with zinc cyanogen compounds. It is obtained by precipitating a solution of 2 parts of zinc sulphate and 1 to 2 parts of green vitriol (according as pale or deep blue is required) by a dilute solution of yellow prussiate.

CHAPTER XXIII.

ULTRAMARINE.

AT the present time blue, green, red and violet pigments come into the market under the name of ultramarine. The green and blue have been commercial articles for about seventy years; the violet and red were introduced about the year 1860.

At first the name ultramarine was restricted to a natural blue pigment obtained from *lapis lazuli*, which was extremely costly. Accounts of the payments of Italian artists, still extant, show the expensive nature of the ultramarine blue used in their paintings. At that time, when artists were compelled themselves to make the majority of their pigments, ultramarine was made in a most laborious manner from *lapis lazuli*, for which incredibly high prices were paid. In order to make the mineral easier to powder the lumps were heated, and, whilst hot, thrown into water. They were then powdered as finely as possible. The powder was mixed with melted resin, and the mixture kneaded under water for a long time. The ultramarine suspended in the water by this crude method of levigation was obtained by allowing the wash waters to settle. Few places are known at which *lapis lazuli* occurs in quantity. It is chiefly obtained in China and Thibet, and, considering the little intercourse between Europe and these distant countries at that time, it is no wonder that the price of ultramarine was fabulously high. One ounce cost about £8—a price which is explained by the small yield of ultramarine from the best *lapis lazuli*. By the most careful work not more than from 2 to 3 per cent. of the mineral was obtained, the residue consisted of foreign minerals.

The enormously high price of this pigment was the stimulus for the endeavours to make it artificially. The attempts are to be regarded not only as completely successful, but it must be allowed that science has gone a considerable step further than nature, since the researches have made it known that there is not only a blue ultramarine, but also a green, and, according to the latest researches, there exist in addition violet, red and white compounds, also to be described as ultramarines. With the discovery of the methods by which ultramarine can be made artificially, the preparation of this pigment from *lapis lazuli* came to an end, and has now but historic interest. The discovery of artificial ultramarine is due to the French chemist Guimet and to the great German chemist Gmelin. Ultramarine was first manufactured in Germany in the year 1828 by A. Köttig, as a branch of the porcelain works at Meissen in Saxony, where the manufacture was continued for about fifty years.

The discovery of the aniline dyes is rightly called a triumph of human intellect. The artificial manufacture of ultramarine deserves the same description in no less degree, although it has not effected so great a revolution as the former.

Attempts to make ultramarine artificially would naturally be based on the analysis of the natural product. The following comparison of the compositions of natural and artificial ultramarines shows how nearly the artificial product approaches the natural :—

NATURAL ULTRAMARINE.

	Clément and Desormes.	Gmelin.
Silica	35·8 per cent.	47·31 per cent.
Alumina	34·8 ,,	22·00 ,,
Soda	23·2 ,,	12·06 ,,
Lime	3·1 ,,	1·55 ,,
Sulphuric acid	— ,,	4·68 ,,
Sulphur	3·1 ,,	0·19 ,,
Water and organic matter	— ,,	12·21 ,,

ARTIFICIAL ULTRAMARINE.

Observer.	Blue.				Green.
	Warrentrapp.	Elfen.	Brunner.	Pohl.	Gentele.
Silica	45·60	40·0	32·54	36·67	} 47·31
Alumina	23·81	29·5	25·25	32·12	
Soda	21·47	23·0	16·91	21·45	39·93 (sodium silicate)
Potash	1·75	—	—	—	3·92
Lime	0·02	—	2·38	—	1·13
Sulphuric Acid	3·83	3·4	—	2·08	—
Sulphur	1·69	4·1	11·63	7·22	6·62
Iron	1·06	1·0	2·25	trace	1·95 (ferric silicate)
Water	—	—	—	—	—
Oxygen	—	—	9·04	0·58	—

I. Szilasi found three samples of green ultramarine to have the following composition :—

Water	2·20	1·20	1·19
Aluminous residue	1·80	1·42	1·41
Silica	16·73	17·18	16·74
Aluminium	15·92	15·87	16·15
Sodium	18·42	18·18	18·10
Sulphur	7·19	6·97	6·85

There are many other analyses in addition to those we have given and agreeing with them, so that there is no doubt as to the composition of ultramarine, but as to the manner in which the elements are grouped nothing is definitely known. Some chemists are of the opinion that the colouring principle of ultramarine is a sulphur compound of iron, whilst others oppose this view and consider that the colour is due to the combination of a double silicate of alumina and soda with an unknown sulphide of sodium. Although no blue or green compound is known of corresponding composition, the majority of chemists incline to the latter view. Experience has shown that the presence of iron in any of the materials used in the manufacture of artificial ultramarine is very dangerous to the success of the operation, and at the least considerably injures the beauty of the product.

Although the manufacture of ultramarine is now very

well known, it cannot be denied that some works produce a pigment of a shade which cannot be obtained by others. These works keep their method very secret, so that it is not possible to say with certainty whether they have introduced a process varying from that commonly known, or whether, by carefully watching the process, they have achieved great technical dexterity in the manufacture of this product. The latter appears the most probable, for in order to obtain a good result, many experiments, and an accurate knowledge of the raw materials, are necessary.

The raw materials used are as follows: pure aluminium silicate, sodium sulphate, soda, sulphur, coal. The aluminium silicate is used in the form of fine china clay or kaolin, sodium sulphate and soda must be used in the anhydrous form, the sulphur is the ordinary commercial substance purified by distillation. Charcoal or coal containing little ash can be used.

Whilst the remaining raw materials are always of similar composition, the china clay from different localities possesses a very varying composition. This substance must be carefully chosen. There is hardly any kaolin which is naturally of sufficient purity to be used without purification; it is well known that the china clay used for porcelain is subjected to a thorough preparation before it is used. Kaolin, like all clays, has been produced by the decomposition of felspar; when the aluminium silicate so formed was able to deposit without foreign admixtures, that mineral was formed which is the purest of all clays and is called kaolin. The more foreign substances are mixed with the aluminium silicate the further is the clay removed from kaolin. The impurities which generally accompany the aluminium silicate are quartz-sand, chalk and ferric oxide. We distinguish accordingly between kaolin, white clay or pipe clay, clay, and lastly marl, a clay containing much chalk.

Even the purest kaolin contains certain impurities, of which quartz-sand is the principal and the least harmful. Before kaolin can be used in the ultramarine manufacture it must always be purified by levigation; it is then ignited at a low temperature and powdered under stamps or in mills. The other materials required are generally produced by the chemical works in a condition of such purity that they can be at once used.

Occasionally Glauber's salt contains iron, which would spoil the shade of the ultramarine. The iron can be easily removed by dissolving the crystallised salt in water, adding a little milk of lime and leaving the liquid for several days, stirring frequently. The lime neutralises every trace of free acid, and at the same time produces an ochre-yellow precipitate of ferric hydrate. The solution of Glauber's salt thus freed from iron is evaporated in reverberatory furnaces, in which the salt is then calcined. This operation is not only simpler than evaporating the solution in iron pans to crystallisation and subjecting the dried salt alone to calcination, but it also guards against fresh contamination by iron, which might be caused by the use of iron evaporating pans.

The sulphur and coal are employed in a soft powder, which is most readily obtained by placing the coarsely powdered materials in rotating drums containing a number of iron balls. By continued rotation the materials are converted to any desired degree of fineness without the production of dust. The powder is then put through fine sieves by which the larger particles are retained.

The proportions in which the raw materials are mixed vary within certain limits; fixed quantities can only be given for a kaolin of definite composition. Definite formulæ are known by which ultramarine is made, but these can only be regarded as approximate. The proportions employed by French manufacturers differ considerably from those usual in

German works. This variation is chiefly due to the difference in the composition of the kaolin employed in the two countries.

From the composition of the different mixtures one conclusion may be drawn with certainty—sufficient sodium is always used to neutralise half the silicic acid of the kaolin and to form some quantity of sodium sulphur compounds. In the successful process of the German makers a portion of the soda unites with the silica during heating. By the action of the coal on the Glauber's salt it is reduced to sodium sulphide, which, since sulphur is present, unites with a further quantity of that element. The sodium sulphur compounds then unite with the silicates of aluminium and sodium to form a green compound, which is converted into blue ultramarine by a further treatment with sulphur in the presence of air.

Instead of using Glauber's salt, which must always be decomposed in the first process, the sodium sulphide may be formed by the action of sulphur on soda in the presence of coal. This procedure is adopted in the French process. The proportions of the mixture used in different works vary. If the kaolin employed is assumed to be bisilicate of alumina—a somewhat arbitrary assumption—the following mixture can be successfully used:—

Anhydrous kaolin	100
,, Glauber's salt	42
,, soda	42
Sulphur	60
Coal	13

In working by the French method the following formula is suitable:—

Anhydrous kaolin	100
,, soda	100
Sulphur	62
Coal	14

These formulæ are not to be regarded as unalterable. In the different works such varied mixtures are used that it may be said with truth that each works has its particular formula for the mixture, the composition of which depends on the nature of the clay used in the works. The composition of the mixtures used in works is kept secret as far as possible. The formulæ given above refer to a clay approximating in composition to the bisilicate of alumina.

It will appear from the description of the manufacture of ultramarine that certain quantities of sodium sulphide are produced. The process is, however, conducted so that sodium sulphide shall be formed; thus the soda and Glauber's salt used in the mixture may be replaced by the sodium sulphide produced in previous operations. The liquors in which the sodium sulphide is contained are evaporated to dryness, and as much of the residue added to the clay as corresponds to the quantity obtained from the usual amounts of soda and Glauber's salt. Assuming that these latter materials are pure, 80 parts of the sodium sulphide residue correspond to 100 parts of soda, and 60 parts to 100 parts of Glauber's salt.

CHAPTER XXIV.

THE MANUFACTURE OF ULTRAMARINE.

WE have thought it necessary to an understanding of the manufacture of ultramarine to discuss the production of this important pigment in some detail, for it is only possible to properly conduct an operation regulated by chemical laws when the processes which take place are accurately known.

The mixture of the materials used in making ultramarine must be most intimate in order that the constituents may act chemically upon one another. In some works the mixing is performed in a very laborious manner—upon a heap of the dried, levigated clay the remaining materials are thrown, and the whole shovelled about till it is completely homogeneous. Naturally, instead of this primitive and costly operation, mechanical mixers may be used, but the mixture can also be effected in a simpler manner, which almost dispenses with the necessity for a mechanical process. Of the substances to be mixed only kaolin and sulphur are insoluble in water; the others and also the by-products of previous processes are easily soluble. The soluble and insoluble constituents of the charge may be mixed in a very simple way by bringing the levigated kaolin in the form of paste, without drying, into a pan, adding the solutions of the salts and the powdered sulphur.

The pans are most conveniently heated by the fire gases from the ultramarine furnaces. After mixing the solutions with the clay to a thick paste the heating is commenced, the solid substances are prevented from sinking to the bottom by continual stirring, the mixture is slowly evaporated, and an extremely intimate and complete incorporation of the constituents occurs. The heating is continued until a dry mass is obtained, which may be used for the second operation without powdering.

The mixture of raw materials is heated for a long time at a clear red heat, or, in some works, at a white heat. The operation demands some care; air must be excluded and the whole mass must acquire a uniform temperature. If both these conditions are not obtained in heating it is very difficult to obtain a product of uniform colour.

In different works different arrangements are used for the calcining process. In the older methods crucibles or dishes of fire-clay were used, of such a shape that the bottom of one crucible formed the lid of the crucible beneath, just as is the case with the saggers used in porcelain kilns. The mixture is closely pressed into the crucibles, of which piles are built in the furnaces. These must be so arranged that each crucible is surrounded by the fire. The furnaces are very similiar to porcelain kilns.

This method of calcining is obviously attended with many drawbacks; a large number of costly crucibles is required, a certain proportion of which is lost at each burning, either by breakage in the fire or in filling. The pressing of the charge into the small crucibles requires much labour, as also the placing of the crucibles in the furnace and their removal; at the end of the nine to ten hours' calcining, it is necessary to wait some time to take the crucibles out of the furnace, or considerable loss will be caused by breakage of the crucibles due to rapid cooling.

For these reasons in most works crucibles have been abandoned in favour of muffles, a large number of which are built in one furnace. The charge is placed in the muffles;

after completion of the calcining it is quickly raked out, and the still hot muffle at once recharged, thus the losses of time and heat are reduced to a minimum. The muffles are generally about 1 metre long, 1 metre wide and 30 to 40 centimetres high. They are made of fire-clay, and, when kept in uninterrupted use, last a long time. The muffles have a small opening at the back which communicates with the furnace, so that the gases evolved have a free outlet without reaching the working place. The front of the muffle is closed by an iron plate in which are small openings, through which the contents can be examined without the entry of much air.

Three muffles are generally built in a furnace, but, by suitable alterations in the construction, a considerably larger number may be heated in one furnace. Whatever the number of muffles, the furnace must be so arranged that the fire can be controlled at will. This is best done by a good damper; all the muffles must be uniformly heated.

J. Curtius recommends cast-iron cylinders, lined with a thin layer of fire-clay. For this purpose a covering of fire-proof cement is used to protect the retorts. This covering, as it is gradually destroyed, may be renewed by smearing over the damaged places with fire-proof cement, mixed with water or some binding material. Aluminium silicate, graphite and coke may also be used. The retorts, a (Fig. 28), project through the wall of the furnace at one end. They rest on fire-proof stone bridges, and can be put into communication with the air by the pipe, m, which can be closed air-tight. Short pipes, k and d, which can be closed, connect the interior of the retort with the cooling and collecting chambers, P and g, the latter of which is connected by r with the vessel, h, where the gaseous products are absorbed. The porous plate, f, prevents the charge in the hinder part of the retort from becoming open and falling to pieces during stoking. In order to obtain a regular heating of the retorts the fire gases passing through the space, q, are led away at two opposite places by flues into the main flue, F. The intimate mixture of finely-powdered materials is introduced into the hinder part of the retort, the porous plate, f, placed in position, the cover, t, closed up air-tight, and the connection with the chamber, P, closed. On heating, the volatile products pass through the porous plate and the pipe, d, into the chamber, g, the more volatile portions, without mixing with the furnace gases, passing direct to the absorption vessel, h, or to a lead chamber, whilst the sulphur which distils over remains in g. When the reaction is finished, by opening

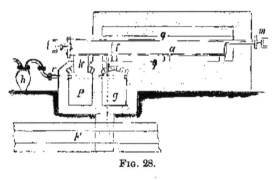

Fig. 28.

the cover, t, and the tube, m, the crude green ultramarine in the retort can be rapidly oxidised to blue by aspirating air at m, or after closing the opening, d, and removing the plate, f, the green ultramarine can be raked into the collecting chamber, P, and there oxidised. The retorts are then immediately re-filled with the mixture and the temperature raised.

The heating of a charge lasts nine to ten hours; the larger the quantity of sodium sulphide in the mixture the shorter is the time required. The mixture is spread out in the muffles in a layer 7 to 9 centimetres thick, and brought to a moderate red heat. When the whole mass is uniformly hot it is more strongly heated; the change of the constituents then becomes visible. The mixture takes at first a brownish colour, and is somewhat similar in appearance to liver of sulphur. The colour soon begins to incline to green, and finally changes to a tolerably pure green, but always has a yellow tinge. Those parts of the hot mass which come into contact with air are more blue, and occasionally change to a pure blue. When crucibles are used the upper layer of the contents is always bluer than the lower, and when crucibles crack during the heating the portions near the cracks always show distinctly a more or less blue colour.

When the calcination is finished the hot mass drawn out of the muffles is at once brought into washing vessels, in which it is treated with water so long as soluble materials are extracted. The compounds dissolved in the water are chiefly sodium sulphide and sulphate and a little alumina dissolved in caustic soda. The presence of a larger quantity of caustic soda in the wash waters indicates too violent heating. In this case the mass will not mix up easily in water; it contains sintered lumps. The washed material is strained, spread out on boards or linen stretched across frames, and completely dried by artificial heat.

At this stage the product may be described as green ultramarine and sold as such, or it may be converted into blue ultramarine. For the latter purpose it is ground to a moderately fine meal, but green ultramarine must be ground to a very fine powder; the finer the powder the brighter the shade.

In order to obtain blue ultramarine from the washed and ground substance, it is subjected for a short time to a moderate red heat with unrestricted access of air. The heating by which the blue colour is produced must be continued until the proper shade is obtained. This operation is conducted in special muffles ; in many works it is customary to sprinkle powdered sulphur upon the heated mass during the roasting. The sulphur burns, forming sulphur dioxide, which escapes into the furnace through an opening at the back of the muffle. The object of the addition of sulphur may be to keep the temperature from rising beyond a certain point, or to prevent a possible reduction of the sulphur compounds in the ultramarine.

The muffles used for finishing the ultramarine are similar to those used in making green ultramarine; they are generally 50 centimetres wide and 100 to 120 centimetres long. A muffle of this size will contain 6 to 18 kilogrammes of green ultramarine. The temperature is gradually raised to a gentle red heat, each muffle being provided in front with an iron plate which prevents the cooling of the contents by the external air. When the charge in the muffles is carefully observed, it is noticed that the blue colour first appears at the surface and the edges, *i.e.*, at those places where the oxygen of the air has unrestricted access. The mass is frequently turned over with an iron rake, in order to expose all parts to the action of the air; samples are frequently taken in order to observe the moment at which the colour has reached its greatest intensity, when the heating is at once stopped. By proper treatment about half an hour is required to convert the charge of a muffle of the above dimensions into ultramarine of the deepest blue. The hot mass is drawn out of the muffles and spread out upon flags so that it may cool quickly. During the cooling the colour is often observed to become considerably darker, which is a sign that the ultramarine has not been heated for a sufficient length of time.

Mechanical arrangements are often used to continually turn over the ultramarine in the muffles, which are then made in the form of cylinders in which the ultramarine is continually turned by a stirrer provided with wings. The whole arrangement is very similar to the apparatus used for roasting coffee.

The ultramarine is then carefully ground to a very fine meal; it is tolerably hard, and granite stones should be used, ordinary stones would be too quickly worn down. It is possible, by the most careful grinding, to convert ultramarine into that condition of fine division which is necessary if it is to be ground into paint; grinding must be followed by levigation. The various qualities and shades brought into the market differ only in regard to the fineness of their particles; they do not vary in chemical nature, for the substance has always the same composition.

It occasionally happens in ultramarine works that a charge does not turn out a bright blue; the product is then sold as inferior quality at a lower price, or is mixed with a larger quantity of good material. In working by a settled process the different shades of ultramarine vary in composition between narrow limits; the difference is caused by variations in the raw materials.

It has been already remarked that there is no agreement as to the composition of the colouring principle of ultramarine; some maintain that it is an iron compound, whilst others regard the iron found in ultramarine as an accidental impurity, which has no connection with the colour. Recently this point appears to have been decided from the results of many most accurate analyses of ultramarine. It has been found that its chemical composition shows the greatest similarity with the mineral nepheline.

Nepheline is a double silicate, its composition is expressed by the following formula: $Na_2O . SiO_2 + Al_2O_3 . 2SiO_2$. By a comparison of the analyses of green and blue ultramarine with that of nepheline, it is seen that blue ultramarine may be regarded as nepheline combined with sodium pentasulphide, whilst green ultramarine is nepheline combined with sodium bisulphide. If this is the correct view, the two species of ultramarine have the following formulæ:—

$$2\{Na_2O.SiO_2 + Al_2O_3.2SiO_2\} + Na_2S_3.$$
$$2\{Na_2O.SiO_2 + Al_2O_3.2SiO_2\} + Na_2S_2.$$

Ultramarine is one of the most permanent pigments. It is not altered by the substances with which a pigment generally comes in contact. Being a sulphur compound, it retains its colour completely in air containing sulphuretted hydrogen. By acids, even weak organic acids such as malic, citric or tartaric, it is rapidly decomposed, sulphuretted hydrogen being evolved, and a greyish-white residue left. It is sometimes found that the sugar used in making lemonade produces a perceptible smell of sulphuretted hydrogen. This is due to ultramarine, which has been added to the sugar to hide its yellow colour.

. From the preceding account of the manufacture of ultramarine it will be understood that there exist only two species of ultramarine, green and blue. In commerce a large number of different shades are found, which are not to be regarded as pure ultramarine; they owe their particular shade to additions. By mixing in a white pigment, such as barytes or starch, the paler shades are obtained; by adding a small quantity of a pure red pigment a colour with a violet tinge is produced. In a similar manner any number of shades of ultramarine can be obtained.

The following account is taken from a new comprehensive work by J. Wunder, on the different ultramarines.

Preparation of Mixtures for Ultramarine.—The following points are to be observed: Chemically pure soda and ammonia soda react with difficulty. The presence of caustic soda in the soda facilitates the formation of sodium sulphide, and gives a finer product. It is also of advantage to sprinkle the soda with a strong solution of sodium sulphide. The more silica the mixture contains the more difficult is the transformation into ultramarine, but the product is deeper and better in shade, and has more resistance to alum and weak acids. The sodium sulphide must react with silica and alumina. When oxygen enters during the burning, silica is re-formed, which with soda and alumina produces slags. In order to exclude oxygen many manufacturers burn with restricted access of air; then the carbon bisulphide and sulphur gases evolved are not burnt, but, together with tarry materials from the coal, form an evil-smelling smoke, which renders the neighbourhood objectionable. The coal consumption in burning is also greater. The varieties rich in silica and stable towards alum are generally dark reddish-blue. In order to obtain pure blue shades from mixtures rich in silica they are burnt in closed crucibles to produce the green, which is powdered and heated with restricted air supply, and the admission of a little steam, to 160° to 180° C., when pure blue or greenish-blue shades are obtained as required. In this way a mixture rich in silica can be burnt to a good blue, containing 69·32 parts of silica to 30·67 parts of alumina, or 1 equivalent of Al_2O_3 to 3·84 equivalents of SiO_2; whilst ultramarine rich in silica, made in the ordinary way, contains at the most 66·7 parts of silica to 33·3 parts of alumina.

Ultramarine Violet was first introduced into commerce about the year 1859 by the Nuremberg Ultramarine Works, under the direction of Leykauf. By the reaction of moist calcium chloride on ultramarine blue in the warm chambers above the furnaces the blue was changed to violet. By the action of air a portion of the moist calcium chloride is decomposed into lime and hydrochloric acid, the latter of which, together with air, reacts on the blue. In 1872 Wunder, intending to apply to the estimation of sulphur in ultramarine blue a method which gives good results with metallic sulphides, passed chlorine over heated ultramarine. The pigment was converted into a brownish-red substance, which on washing combined with water and turned violet. The red substance is not produced at every temperature; 300° C. is the most favourable. On washing sodium chloride is dissolved; the washed violet is free from chlorine, but has taken up water, which is expelled on heating, when the violet turns to blue. A similar brownish-red body is formed when sulphur trioxide is led over heated ultramarine blue, and when sulphur chloride acts on warm ultramarine blue. When the brownish-red chlorine compound is introduced into absolute alcohol, a reddish-violet compound is produced, containing organic matter. Ethyl chloride is formed at the same time. When ammonia gas is passed over the heated chlorine compound it is absorbed; a violet ultramarine is produced, from which the ammonia is not removed by washing with water, but only by heating nearly to redness, or, better, by fusing with caustic alkalis. Aniline combines in a similar manner. In order to obtain a good violet the brownish-red chlorine compound must not be formed; the violet is produced by leading chlorine and steam over ultramarine blue at 160° to 200° C. Blue rich in silica is most suitable for transformation into violet. The violet so obtained is not decomposed by lime.

C. Mahla produced the chlorine in the blue itself by the reaction: $NH_4Cl + 2NH_4NO_3 = 6H_2O + 5N + Cl$. By heating a mixture of ultramarine blue with ammonium chloride and nitrate in crucibles at 200° C. a fine violet is produced. It contains ammonia, which cannot be removed by washing, but only by strongly heating or heating with concentrated alkalis. This violet is decomposed by the prolonged action of lime, in three days it is changed to a grey green. In the course of the manufacture it was noticed that less nitrate was required in the mixture according to the length of time it was exposed in porous crucibles to the action of air and heat; by heating for a sufficient length of time with access of air, a good violet is obtained with ammonium chloride alone. In the manufacture of violet by means of moist chlorine hydrochloric acid is formed, which is also the active material in its formation by means of ammonium chloride. It was therefore to be expected that chlorine gas might be replaced by hydrochloric acid, if it were accompanied by sufficient air. This has been found to be the case. The temperature must be maintained between 180° and 230° C. Below 150° C. the blue is decomposed by moist hydrochloric acid gas; at temperatures above 230° C. it is unchanged. Hydrochloric acid gas and air, without steam, give with ultramarine blue on heating a brownish red substance similar to that produced by dry chlorine; it also is changed to violet by washing with water. The violet obtained by means of moist chlorine differs from the blue from which it was formed, in that the latter has lost one-sixth of its sodium and combined with water and much oxygen. The following is the approximate formula for a blue:—

$$Na_6Al_4Si_6S_4O_{21},$$

and for the violet obtained from it

$$Na_5HAl_4Si_6S_4O_{24} + H_2O.$$

The violet contains much thiosulphate. If it is decomposed by nitric acid and silver nitrate added to the filtered solution a precipitate is obtained which changes in colour from white to yellow, orange and brown, just as the precipitate given by silver nitrate in solutions of thiosulphates containing nitric acid.

Chlorine and Steam Process.—Chlorine does not attack iron at 150° to 250° C. The reaction is carried out in heated iron boxes 1 metre wide, 2 metres long and 65 centimetres high. The blue is spread out in a layer 2 centimetres thick on earthenware plates, which stand one above another at a distance of 5 centimetres, each supported on three feet. The plates are introduced by means of iron tongs through large openings in the top of the box; after filling iron plates are screwed on to the openings, through them pass sheet-iron tubes reaching to the bottom, in which thermometers can be lowered on wires. The iron boxes stand in heated chambers, which have openings corresponding with the openings in the boxes, and shut off from the interior of the furnace. Chlorine and steam are led into the iron boxes at both ends through lead tubes reaching to the bottom; the parts of these tubes in the furnace and the iron boxes are protected by wide sheet-iron tubes, the space between being filled with clay. The gases evolved, hydrochloric acid, sulphur chloride and steam, pass from the covers through earthenware pipes into boxes filled with limestone, upon which water drops, and thence to the chimney. After filling, the boxes are heated to 280° C. and steam introduced to remove sulphur; they are then allowed to cool to 160° C., when chlorine and steam are led in for three hours. The violet is then finished, and the residual gases are blown out by a fan.

Hydrochloric Acid and Air Process.—The operation is conducted in the same iron boxes, upon the bottoms of which, beneath the openings, are earthenware dishes into which hydrochloric acid can be poured through earthenware pipes. The chimney draught must be strong enough to draw sufficient air through these pipes into the boxes. After the temperature has been maintained for seven hours at 220° to 230° C., hydrochloric acid being poured in from time to time, the blue is changed to a dull violet, which becomes brighter when hydrochloric acid is repeatedly added at diminishing temperatures, 210°, 200°, 180° and 160° C. More recently the iron boxes have been replaced by stone chests or by chambers above the ultramarine furnaces, the temperature in this case being kept at about 200° C.

Ammonium Chloride Process.—A mixture of ultramarine blue with 5 per cent. of ammonium chloride is heated during fourteen days in porous crucibles placed in the upper chambers of the ultramarine furnaces, when the contents of the crucibles become a handsome violet throughout. If sodium nitrate is used together with ammonium chloride, the violet is formed in a much shorter time. After washing, the violet still contains nitrogen; on ignition or heating with strong alkalis it loses ammonia. Unfortunately, this fine violet is decomposed by the continued action of moist slaked lime. The violets made by the first two processes absorb ammonia when this gas is led over them at 180° to 200° C., and it cannot be removed by washing.

Pale Blue Ultramarine.—If the violet is heated in hydrogen at 280° to 290° C., it is converted into a pure, bright, pale blue. This has an absorption spectrum in which the red is not absorbed, but appears more brilliant than in the spectrum of ultramarine rich in alumina. Pale blue is turned violet blue by heating at 300° C., and at a red heat a dull blue. It is not yet made on a commercial scale, but on account of its great purity of shade it appears to be valuable for many purposes; perhaps it may replace alumina cobalt blue. The composition of pale blue is:—

	Calculated for		Found	
Na_2	12·4	per cent.	11·9	per cent.
H_5	0·54	,,	0·62	,,
Al_4	11·7	,,	13·1	,,
Si_6	18·2	,,	19·7	,,
S_4	13·9	,,	12·7	,,
O_{26}	43·3	,,	42·0	,, (by difference).

By a comparison of the composition of the violet and pale blue ultramarines, it is seen that the chief difference is an increase of hydrogen in the latter.

Ultramarine Red.—Since ultramarine violet increases in brightness and redness of shade in the air, Wunder erroneously believed that this was due to oxidation, and that, consequently, the violet could be converted into a red by oxidising agents. Nitric acid vapours led over ultramarine violet at 170° to 200° C. do not act upon it, but where drops of nitric acid are spirted over, the violet is changed to red. Wunder then reduced the temperature to 135° to 145° C. and obtained the first ultramarine red. Iron is attacked by nitric acid at lower temperatures, but not at 135° C.; the iron boxes previously described could therefore be used. It was afterwards found that at a sufficiently low temperature hydrochloric acid gas converts ultramarine violet into red. The iron boxes cannot be used for this operation, as they are attacked at the temperature; the stone chests are used instead. Other acids also act on ultramarine violet; boric acid gives a reddish violet.

The violet is spread out on the dishes standing on three feet mentioned before, and heated to 128° to 132° C. At higher temperatures the violet is unaltered, whilst below 100° C. it is decomposed. The hydrochloric acid is poured in from time to time through earthenware tubes into dishes in which it evaporates.

A mixture of red and blue would appear violet, but would behave towards reagents in a different manner to real ultramarine violet. Ultramarine blue is decomposed at 128° to 132° C. by hydrochloric acid to a gelatinous mass, whilst at this temperature ultramarine violet is changed into a bright red. From the blue no violet can be obtained by nitric acid, but the violet gives a red at 135° to 145° C. Analysis would also indicate the difference. Ultramarine red has the following composition:—

Calculated for			Found.		
Na_2	7·9	per cent.	8·1	per cent.	
H_3	0·57	,,	0·72	,,	
Al_4	12·3	,,	13·3	,,	
Si_4	19·1	,,	19·3	,,	
S_4	14·6	,,	15·2	,,	
O_{25}	45·6	,,	43·4	,,	(by difference).

It appears that in the red two more equivalents of sodium have been replaced by hydrogen. The violet is apparently a sodium salt of which the red is the acid. The violet made by means of ammonium chloride is also converted by hydrochloric acid gas at 128° to 132° C. into a handsome red containing nitrogen, and the red is changed by hydrogen at 280° to 290° C. into a lighter pale blue.

CHAPTER XXV.

BLUE COPPER PIGMENTS.

Bremen Blue and Green.—The pigment known as Bremen blue or Bremen green consists of copper hydroxide. It possesses all the disadvantages of copper compounds; it is very sensitive towards sulphuretted hydrogen and sulphur dioxide; in contact with air containing the former gas it is quickly turned black. It has also other disadvantages; it cannot be used in oil, for when ground in oil it is quickly discoloured, on account of the formation of copper oleate. If a wall which has not completely dried be covered with Bremen green, in a short time it will be covered with spots. In spite of this small stability, Bremen green is still largely used by artists on account of its low price, although it could be replaced by other pigments almost as cheap and considerably more durable.

Copper hydroxide comes into the market under many names. The pigments known as blue verditer, lime blue and mountain blue contain essentially copper hydroxide. The preparation of pure copper hydroxide on the large scale will be first described. Copper sulphate is the usual raw material. In order to obtain a colour of a pure blue shade it must be free from iron, the presence of which would result in the production of a discoloured precipitate.

A fairly concentrated solution of copper sulphate is warmed in a pan to about 30° C.; weak caustic potash solution is added until the liquid is slightly alkaline. A green precipitate separates, which consists of copper sulphate and hydroxide. In order to remove the copper sulphate from the precipitate, and impart to it the correct bluish-green colour, it is left on a strainer until it has acquired a pasty consistency, when it is brought into a tub and mixed with weak caustic potash solution. Care must be taken that this potash solution is not too concentrated, or it will withdraw water from the copper hydroxide and produce black copper oxide. In order to prevent this result, which would render the whole mass useless, a portion of the precipitate is tested with the caustic potash solution. If it is blackened, the solution must be diluted with the necessary quantity of water before it is used. Instead of diluting with water, potassium carbonate solution is often added. The addition of the caustic liquor is continued until the colour of a test portion of the precipitate is not deepened by further additions. When this point is reached the colour must be at once withdrawn from the action of the alkali. It is brought into a large tub filled with water, in which it is most carefully washed with many waters.

Basic copper chloride may also be used as the raw material for this pigment. It is obtained by treating scrap copper with 60 per cent. of common salt and 30 per cent. of sulphuric acid. The pale green basic chloride is then treated with caustic potash solution. This method is now rarely used; copper sulphate is considerably lower in price than formerly and copper hydroxide is prepared from it with the least trouble.

Neuberg Blue is a mixture of copper blue with Chinese blue. The larger the quantity of the latter component the deeper the colour. Neuberg blue can be more easily ground with oil than pure copper blue. It should not be used as a distemper colour, for the Chinese blue is decomposed by the action of lime, and the colour will quickly be spoiled by the separated ferric oxide.

Lime Blue is ordinary copper blue, mixed with varying quantities of gypsum. It is paler than pure copper blue; different shades are obtained by increasing the quantity of gypsum.

This pigment is made by dissolving 15 parts of copper sulphate in 1,300 parts of water, adding a strong solution of 121 parts of ammonium chloride, and then milk of lime prepared from 30 parts of quicklime. The precipitate is well washed, strained, ground, and finally dried.

Lime blue is brought into the market in several forms; in cubes obtained by cutting up the stiff paste and then completely drying, and in irregular lumps or a coarse powder. The pigments containing copper hydroxide as their colouring principle are used less as artists' colours than for distempering.

Payen's Mountain Blue is a mixture of copper hydroxide with varying quantities of calcium carbonate. It is prepared by adding calcium chloride to a solution of copper sulphate and stirring in dilute milk of lime. A green precipitate of basic copper chloride is formed; potassium carbonate solution mixed with milk of lime is then added. The latter mixture contains caustic potash, which decomposes the copper chloride, copper hydroxide and calcium carbonate being simultaneously precipitated. Different shades may be obtained; by increasing the quantity of lime the product is paler. This pigment commonly comes into the market in the paste form. When it is dried the operation must be conducted at a low temperature; copper hydroxide is converted into copper oxide at relatively low temperatures.

Oil Blue.—The pigment known under this name consists of copper sulphide (CuS). It can be made in several ways. The simplest process, which also gives a good product, is here described.

Sulphur is boiled in a glass flask with a long neck. When

the heavy reddish-brown vapours begin to fill the neck, copper turnings or wire free from oxide are thrown in. The copper burns in the sulphur vapour with a red flame to cupric sulphide. When it has once commenced to burn, the flask requires little heating, so much heat is produced by the combustion of the copper that the sulphur continues to boil. The introduction of copper is continued until only a small excess of sulphur is left, the flask is then closed and allowed to cool. When the contents have reached the temperature of the air the flask is broken, and the mass boiled with caustic potash solution in order to remove excess of sulphur.

The product obtained by this process has a handsome violet-blue colour, which must be shown by the whole mass. If portions taken out of the flask are black, an excess of copper has been used. In order to improve the colour of a faulty batch, it is mixed with sulphur and quickly heated to the boiling point of sulphur, air being excluded. The excess of sulphur is then removed by boiling with caustic potash solution.

Oil blue is a handsome but not very durable pigment. When used in varnish, which protects it from the action of air, it is fairly permanent.

Copper Hydroxide.—By precipitating carefully a solution of a copper salt with caustic soda, a blue precipitate of copper hydroxide is obtained, which, though not distinguished by any particular beauty, is adapted for use as a distemper colour and for similar purposes. The operation is conducted at ordinary temperatures; on account of its voluminous nature the pale blue precipitate requires a long time for washing. Hot solutions must not be used; at a temperature near the boiling point of water copper hydroxide is completely decomposed into black copper oxide.

CHAPTER XXVI.

BLUE COBALT PIGMENTS.

Cobalt Blue, Thénard's Blue, Cobalt Ultramarine, King's Blue, Leyden Blue.—The pigment known under these names is a compound of cobalt oxide with alumina, and is prepared in a similar manner to Rinmann's green. Of all the blue pigments used in painting, cobalt blue is the most handsome and most durable. It is completely unaltered by the atmosphere. The most powerful chemical reagents have but a slight action upon it, and their action is less the higher the temperature used in the manufacture. Cobalt blue is made by precipitating mixed solutions of a cobalt salt and alum, washing, drying, and igniting the precipitate. A product of pure blue colour is only obtained when alum absolutely free from iron is used. If but very small quantities of iron are present, the red ferric oxide has a very considerable influence on the shade of the product.

The depth of colour varies according to the proportion of cobalt salt used. The variation is not so considerable as in the case of cobalt green; as a rule, 50 to 100 parts of cobalt sulphate or nitrate are used to 100 parts of alum. The temperature at which the precipitate is ignited must be much higher than that used for cobalt green; it should be raised to a white heat. The crucible should be provided with a well-fitting lid—the fire gases damage the shade of the colour.

When cobalt sulphate is ignited with ammonia alum, cobalt blue is obtained; but this method is not to be recommended. The ignition must be of long duration to completely drive off the sulphuric acid. The author has made a cobalt blue of remarkable beauty in the following manner: Alumina obtained by precipitating electrolytically an alum solution (page 49) was mixed with a solution of pure cobalt chloride to a thin pulp, which was quickly dried in a shallow porcelain dish. The mixture, whilst still hot, was pressed into a crucible placed in a larger one filled with magnesia; the crucible was heated to a good white heat, at which it was maintained for an hour, when it was allowed to cool. The cobalt blue made by this method has the purest blue colour, being without that admixture of red which is not rarely to be perceived in this pigment. To avoid the appearance of this red tint, certain additions are made; a small quantity of a zinc salt is most frequently employed, according to the quantity of which more or less pale shades are produced.

Cæruleum is a pale blue pigment consisting of a compound of the oxides of cobalt and tin. It is made by converting tin by the action of nitric acid into white stannic oxide, adding a solution of cobalt nitrate and evaporating to dryness in a porcelain dish whilst stirring; whilst still warm the powder is strongly ignited. In cæruleum the chemical combination is less firm than in the compounds of cobalt oxide with zinc oxide or alumina, strong hydrochloric and dilute sulphuric acids decompose it, continued treatment with nitric acid extracts all the cobalt. Cæruleum resists completely all atmospheric influences.

The circumstance that all the cobalt in this compound may be removed by treatment with nitric acid leads to the conclusion that the pigments composed of cobalt oxide and another oxide are not chemical compounds of the two oxides, but that the alumina, zinc oxide and tin oxide serve simply as bases for the blue modification of cobalt oxide. That a chemical compound cannot be present follows from the fact that the quantities of the oxides uniting together may be very considerably altered and a coloured product still obtained. The author has shown in 1868 that the blue modification of cobalt oxide is obtained by heating the red. The change of the red colour of cobalt chloride into the blue is not caused, as was formerly thought, by loss of water, but by the transformation of the cobalt oxide contained in the compound from the red into the blue modification. Strong bases are able to retain the cobalt oxide in the blue form, and strongly basic oxides are therefore used in the manufacture of this series of pigments.

Cobalt Zinc Phosphate.—This pigment is made by precipitating a solution of zinc sulphate, free from iron, by sodium phosphate, adding a solution of cobalt sulphate and again sodium phosphate so long as a precipitate is produced. Excess of sodium phosphate is to be avoided, otherwise a considerable quantity of cobalt remains unprecipitated. In the directions given by Gentele it is stated that the sodium phosphate must be in excess, and that the liquid above the precipitate must possess a reddish colour. The author's experiments have shown that a pigment in every respect equal to that produced by Gentele's process is obtained when all the cobalt is precipitated. The precipitate, which consists of variable quantities of zinc and cobalt phosphates, is dried and ignited, when it acquires a very deep blue colour.

CHAPTER XXVII.

SMALTS.

This blue pigment is a very finely powdered glass, coloured by cobalt oxide, according to the quantity of which products of very different shades are obtained. If a very small amount of cobalt oxide is used the smalts has a beautiful pale blue colour, but by the use of larger quantities an almost black mass can be obtained.

Although the proportion of cobalt oxide contained in smalts is so variable, yet it is probably present as a compound of constant composition—the cobalt potassium double silicate of the formula $CoO \cdot 2SiO_2 + K_2O \cdot 2SiO_2$. The numerous published analyses of smalts lead to this conclusion. The small quantities of other metallic oxides which are found in smalts are to be regarded as chance impurities. Being a glass free from lime smalts is easily decomposed; in this respect it shows great similarity with water-glass, which is completely soluble in boiling water.

Smalts comes into commerce in the form of powder. In the "Blue Colour Works" in which smalts is made a number of qualities are distinguished; these are separated according to the fineness of the powder. Smalts of good quality is composed of particles of the same size and has a pure sky-blue colour. A shade inclining to red is generally to be ascribed to the presence of iron; material of this nature is of little value.

Smalts is not easily attacked by chemical reagents. It has

been above stated to belong to the easily decomposed glasses, and it is attacked by reagents which ordinary glass completely resists. Finely ground smalts is strongly attacked by boiling sulphuric acid, and, after fusion with soda, is soluble in acids. Water has some action upon smalts; by the long continued action of a large quantity the shade is injured, the colour acquires a greenish tinge.

In the manufacture of smalts there are three principal operations:—

1. Preparation of the charge.
2. Fusion of the charge.
3. Grinding the fused mass.

These processes are principally mechanical. Only as regards the composition of the charge is chemical knowledge required.

1. Preparation of the Charge.—The cobalt is supplied by substances of different origin—as a rule, roasted speiss cobalt and cobalt glance are used; occasionally ores are employed without previous preparation. When pure cobalt salts (sulphate and nitrate) are used, a product of very fine shade is produced, but in practice these salts are not applicable on account of their cost. There remain therefore only the compounds of cobalt occurring in nature. On account of the foreign substances they contain, these require a special treatment before they can be used in the manufacture of smalts; the treatment varies according to the other elements present in the cobalt ores. Ores containing bismuth are first treated in order to obtain this valuable metal; ores containing arsenic are roasted in furnaces of special construction, which are connected with a series of stone or wooden passages in order to condense the arsenic trioxide completely.

If ores are employed which contain only a small quantity of cobalt, the same quantity of glass is melted time after time with fresh quantities of the poor ore. At each operation cobalt is taken up. In place of glass a mixture of potash and very fine quartz sand may be used with advantage. It produces that species of glass which most easily takes up cobalt oxide. The quartz sand is generally made by igniting flint and quenching in water to facilitate the grinding; it is then powdered. River sand is generally so impure that it would produce a brown or green coloured glass, which, when fused with cobalt ores, would give smalts of a bad shade.

2. **Fusion of the Charge.**—This process is very similar to the fusion of ordinary glass. The materials, in very fine powder, are carefully mixed and then "fritted," that is, heated until the mass sinters together without melting. The real fusion is accomplished in glass furnaces. The fritted mass is fused in small "glass pots" made of fire-clay, in a furnace similar in principle to the ordinary glass furnace. The glass pots are made rather small, to hold about 50 kilogrammes, in order to obtain as homogeneous a product as possible. They are placed to the number of six in a furnace heated by coal, in which a very high temperature can be obtained. This is essential, because the charge has a high melting point and must be made quite fluid in order that impurities may settle.

The furnace must be slowly heated at first to prevent the pots from cracking. When they are once heated the temperature is raised to fuse the charge, which is repeatedly stirred so that the residues may sink to the bottom. These residues consist chiefly of metallic sulphides and arsenides. After six to seven hours' fusion the stirring is discontinued, and the temperature raised for a short time to the highest it is possible to attain, so that the smalts may be as fluid as possible. When this is the case it can be drawn out into long thin threads. The melted glass is then taken out and poured into cold water. The rapid cooling makes it extremely brittle, and considerably facilitates the powdering.

When new pots are used for the fusion, their walls are considerably attacked by the alkalis contained in the glass. In order to prevent this a small quantity of common smalts is fused in them before the introduction of the real charge; the inner surface then becomes covered with a protective glaze.

3. Grinding the Fused Mass. — The smalts is next powdered under stamps. These only differ from the ordinary construction in that both stamp and trough are made of granite. The coarse powder from the stamps then goes to mills of ordinary construction, in which it is ground as finely as possible in a current of water. A blue liquid runs from the mills into tubs. The coarser particles which settle in these vessels constitute the article known commercially as strewing smalts, which is of small value. The quantity of this quality is fairly considerable; it cannot all be sold; part is fused in a later operation with a fresh quantity of the fritted mixture. The liquid remaining in the tubs then gives the different qualities of smalts. After having deposited the coarsest particles in the first vessel it is brought into another, in which it stays only a few minutes, when it is run off into another vessel, in which also it remains but a short time. The liquid goes in this manner through from three to five vessels, and in the process deposits all the coarser particles. The oftener the liquid is brought into fresh vessels the finer are the particles still suspended in it, and the longer it should remain before being drawn off into another.

When the liquid has run through three or four vessels, as the case may be, it still retains the finest particles in suspension. It is now run into a tank, in which it remains until it is perfectly clear and has deposited all the blue. The deposit in this tank has a dirty bluish-grey colour; it is utilised by being added to a later fusion. It would be expected that this last deposit would have the best shade, since it consists of smalts in the finest state of division, which would no doubt be the case if the smalts remained unchanged. Smalts is, however, a glass which is rather easily decomposed. In grinding, the smalts comes in contact with large quantities of water, and in the succeeding levigation the water decomposes it; the final residue is thus produced. In view of the easy decomposition of smalts by water, it is necessary to carry out all processes in which it is in contact with water as rapidly as possible, in order to obtain the largest yield of smalts.

The smalts deposited in the successive vessels have a different degree of fineness, and are sold under particular marks. The Saxony works, which produce the greater part of the smalts used in Germany, distinguish the qualities by letters. The better qualities are distinguished by the same letter, the fineness of the quality being indicated by the repetition of this letter. Thus FFFE, FFE, and FE indicate qualities which are of lower value the less often the letter F occurs in the mark. The qualities following those marked F are distinguished by M; the lowest quality is marked OE. F indicates fine; M, medium; and O, ordinary.

The qualities generally designated smalts are those deposited in the third and fourth vessels; they are also known as king's blue or azure blue.

The processes used in different works apparently vary. The essentials of the processes are, however, as they have just been described; the time during which the liquid is left in the various depositing vessels is different.

Smalts, being a glass, well resists the action of the weather, although on account of the absence of lime it has not the great chemical indifference of ordinary glass. Smalts is not altered by atmospheric influences, but from analogy with the behaviour of water glass, to which smalts is closely related in chemical composition, we should conclude that, although smalts resists the action of the air for a tolerably long time, it cannot be regarded as absolutely permanent; in the course of time potassium carbonate and silica will be formed.

Since the introduction of the manufacture of ultramarine, the use of smalts has considerably decreased. Ultramarine is incomparably the more handsome pigment, but it cannot be exposed to high temperatures without undergoing change of shade. For purposes in which the pigment is subjected to a high temperature, as in porcelain painting or earthware glazing, smalts has maintained its position. Smalts is also used in fresco painting, for colouring wax and tinting writing paper. For the latter purpose it should be rejected. It has no influence on the durability of the paper, but it spoils steel pens. Any one who has written for a long time on paper coloured with smalts must have observed this; the best steel pens are soon worn out by such paper. The very fine splinters of glass present in large number in the paper act upon the steel as a very fine but very hard file, and quickly wear it down.

When smalts are used for painting earthenware, the outlines of the design are quite obliterated, the cause of which is that smalts, being itself a glass, fuses with the glaze (also a glass) of the vessel. When clean outlines are required smalts should not be used; some other cobalt pigment resisting heat should be substituted.

In addition to the above blue pigments which are in common use, there are two others to be mentioned, rarely used and very costly—tungsten blue and molybdenum blue.

Tungsten Blue is made by adding excess of ammonium chloride solution to a hot solution of potassium tungstate, collecting the precipitate resulting when the liquid has completely cooled, drying and heating to redness in a crucible, through the cover of which a porcelain tube reaches to the bottom. When the crucible begins to glow, hydrogen is passed through the tube; the crucible is then heated to a full red heat for fifteen minutes, the current of hydrogen being continued. The pigment obtained in this way is a deep blue powder of velvety appearance.

Tessié du Motay's Blue.—The following is the method of preparation: 10 parts of sodium tungstate, 8 parts of tin crystals, 5 parts of yellow prussiate, and 1 part of ferric chloride are dissolved separately and the solutions mixed. The precipitate is washed and exposed to light in thin layers. The blue colour develops in the course of several days. According to the discoverer, this pigment consists of a compound of tungsten oxide with a double cyanide of iron and tin; in physical properties it is similar to good Prussian blue, but is distinguished by greater durability on exposure to light and also by a much higher cost. Up to the present this pigment has not been made commercially.

Molybdenum Blue.—This fine but costly pigment is obtained by mixing a solution of sodium molybdate with a solution of stannous chloride. A blue precipitate is formed, which, after washing and drying, may be used as an artists' colour. It is distinguished by great durability. A good shade of molybdenum blue can be obtained with greater certainty by adding finely-powdered tin and a few drops of hydrochloric acid to a solution of pure molybdic acid. In composition it is a mixture of stannic molybdate and the blue modification of molybdenum oxide.

CHAPTER XXVIII.

GREEN MINERAL PIGMENTS—GREEN COPPER PIGMENTS.

Copper Carbonate.—In nature two salts occur containing copper, carbonic acid, and water in different proportions—one, the mineral malachite, has a fine emerald green colour; the other, azurite, is sky blue. Both minerals are ground and levigated, and brought into commerce as mountain green and mountain blue.

Compounds of copper with carbonic acid, made artificially, are used under different names as pigments. The terms mountain blue and mineral blue are applied indiscriminately to various copper compounds; as mountain blue, or Brunswick green, pigments come into commerce containing varying quantities of copper carbonate (made by precipitating a copper salt with soda) mixed with barytes, zinc white, or another white pigment, in order to obtain paler and deeper shades.

The poisonous Scheele's green is also given the same name as that under which copper carbonate is sold, *viz.*, mineral green, so that the composition of this colour can only be decided by analysis.

A green pigment, similar in composition to malachite, is obtained by precipitating copper sulphate with soda in the cold and washing quickly. If the precipitate is left in contact with the liquid, its colour changes and it is converted into a blue mass similar in properties to ground malachite. A green precipitate is obtained with more certainty by precipitating a boiling solution of copper sulphate with soda solution.

Both green and blue copper carbonate are pigments of little depth; as a rule ground and levigated malachite and azurite have a greater intensity of colour than the artificial pigments.

Copper Arsenite.—The compounds of arsenious acid with copper are among the most handsome and bright colours which exist. It is thus to be regretted that these pigments are disappearing more and more, and indeed with every reason, from the pigments permitted by law. In the end they must fall into complete disuse. It is a matter of common report that these pigments, and particularly emerald green, have frequently caused arsenical poisoning. Numbers of people have been repeatedly made ill by the dust of emerald green fixed on light fabrics, such as are used for ball dresses, simply by albumin or dextrine, and brought into the air by the movements of the dancers. It has been further shown that the air of rooms papered with hangings coloured by emerald green contains small quantities of the most poisonous of all arsenic compounds, arseniuretted hydrogen.

Although it is in the interest of colour makers to produce the brightest possible colours, yet, on account of their dangerous properties, pigments containing arsenic should be quite excluded. In several countries their use is forbidden by law. Since there are several pigments which can compete in brightness with the arsenic colours, it is to be hoped that the latter will shortly disappear from commerce.

Scheele's Green, Swedish Green.—A solution of 32 parts of potash is boiled with 11 parts of white arsenic (arsenious acid) until the latter is completely dissolved. This solution of potassium arsenite is mixed with a hot solution of 32 parts of copper sulphate free from iron. The beautiful green precipitate is washed with hot water and dried. These directions for the preparation of copper arsenite are due to Scheele, the discoverer of the pigment. It can be prepared by several other methods.

When white arsenic and copper sulphate are dissolved together, and caustic soda added to the boiling solution, Scheele's green is precipitated. When made by this method the pigment must be very carefully washed; if the excess of caustic soda is not completely removed it is very hard when dry, and is difficult to powder. In order to obtain darker greens the quantity of copper sulphate is increased.

A particularly bright green is obtained by dissolving copper sulphate in water, with the addition of 12 to 15 per cent. of white arsenic, and precipitating by a solution of potassium zincate, which is obtained by boiling caustic potash solution with zinc filings, hydrogen being evolved.

The shade of Scheele's green depends on the relative quantities of arsenious acid and copper sulphate employed. If the arsenious acid predominates, a pale product is obtained; when excess of copper sulphate is used, the colour is pure green. The pigment made according to the foregoing directions possesses only small covering power. A product better in this respect is obtained by precipitating a solution of 100 parts of copper sulphate by a solution of 90 parts of potassium carbonate, in which 66 parts of arsenious acid have been dissolved.

Scheele's green is a bright green pigment, which resists the action of the atmosphere tolerably well. It is easily decomposed by dilute acids; when strongly heated arsenious acid is set free, and copper arsenate and arsenide are produced. Copper arsenite is the essential constituent of the pigments sold under the name of mineral green, which is generally a mixture of copper arsenite with varying quantities of mountain green.

Brunswick Green, Green Verditer, is a mixture of copper arsenite, copper hydroxide and gypsum. It is prepared by dissolving 100 parts of copper sulphate in water, adding a solution of $\frac{1}{4}$ part of arsenious acid and 10 parts of anhydrous potassium carbonate, and finally precipitating with milk of lime made from 20 parts of quicklime. The pale to bluish green precipitate is washed and pressed into flat cakes, which are brought into the market and used as lime colours. This pigment is not adapted for use in oil; when applied in that medium it darkens considerably.

Neuwied Green is made in almost exactly the same manner as Brunswick green, the chief difference being that a larger quantity of arsenic is used, and thus a product obtained inclining more to green than to blue. As a rule 2 to $2\frac{1}{2}$ parts of arsenious acid are used to 100 parts of copper sulphate. The colours found in commerce under this name very often consist of mixtures of emerald green and finely ground barytes.

Copper Oxychloride was at one time largely used. At present it is scarcely employed; it has been replaced by handsomer and cheaper pigments. Copper oxychloride is made by exposing a wet mixture of copper, copper sulphate and common salt to the action of the air. To 111 parts of copper sulphate 112·5 parts of copper and the same quantity of common salt are employed. The mixture is made into heaps, which are wetted and from time to time shovelled about, so that the air may come in contact with all portions of the mass, and the copper chloride be converted into basic chloride. It is advisable to allow the heaps to dry frequently after they have been turned over, so that the air may more readily penetrate into the interior. Copper oxychloride has a pale green colour of little brilliance. On this account it is hardly ever used now; it forms the raw material for the preparation of Bremen green.

CHAPTER XXIX.

EMERALD GREEN.

THERE are many versions as to the discovery of this pigment. According to one it was first made in Schweinfurt by Russ and Sattler; another version gives it that Mitis of Vienna first made it on the large scale. It is possible, as has been the case with many chemical products, that this pigment was simultaneously discovered by both. The method of making the pigment was at first a secret possessed by few. Not until the publication of a method by Liebig in 1822 was the preparation of emerald green generally known. In recent times the industry has supplied other colours which are less poisonous. The time should not be far distant in which this handsome but highly poisonous pigment shall no longer be found in commerce.

The circumstance that the poisonous nature of the pigment and cautions as to its use were given great publicity, brought about the discovery of the countless names under which it has been sold. It was endeavoured by re-naming the pigment to pass it off as a different substance. All the pigments which have been sold under the names of Mitis, moss, patent, new, king's green, etc., are either pure emerald green or mixtures of it with barytes or gypsum, which additions were intended to alter the shade of the green and make it appear to the buyer as a new pigment.

In chemical composition emerald green is a compound

of copper arsenite and copper acetate. It has the following formula:—

$$3(CuO.As_2O_3) + Cu(C_2H_3O_2)_2.$$

This compound rarely comes into commerce in the pure state; it is generally mixed with chrome yellow or lead sulphate, which are added with the object of producing shades between yellowish green and dark green. Emerald green is a compound of constant composition. When made by different processes it has the same properties. Under the microscope it is seen to consist of crystals; when the pure substance is ground it becomes paler, the intensity of the colour being diminished by the breakage of the crystals. The colour of emerald green is not altered by artificial light. This is a most valuable property; the majority of green pigments appear yellowish-green in artificial light.

Emerald green can be made on the large scale by two processes; these differ according to the copper salt used, which is either the acetate (verdigris) or the sulphate. The latter is by far cheaper than verdigris, which is now seldom used for this purpose.

Manufacture of Emerald Green from Verdigris.—By boiling for several hours, 70 parts of ordinary verdigris are dissolved in water; in a second vessel 100 parts of arsenious acid are dissolved in 1,500 parts of hot water; the verdigris solution is filtered through a cloth into a larger wooden vessel and mixed with two-thirds of the arsenic solution. The mixture is well stirred and allowed to stand for some hours. A precipitate of dull green colour is formed. This consists of copper arsenite. After three to four hours the remaining one-third of the arsenic solution is added; the precipitate then changes into the double salt of which emerald green consists.

The use of such dilute solutions as are given in these

directions is essential when a product of particularly good colour is required. Emerald green is a crystalline substance which is deeper in colour in proportion to the size of its crystals. Large crystals can only be formed from very dilute solutions; when strong solutions are used, the emerald green has never the deep colour possessed by the product of dilute solutions. In the place of ordinary (basic verdigris) the so-called distilled verdigris (neutral copper acetate) can be used, of which seventy-six parts are required for the quantity of arsenious acid given. The emerald green made from "distilled" verdigris is of a yet deeper colour than that made from ordinary verdigris, but is considerably dearer. Such emerald green is commercially described, very incorrectly, as distilled emerald green.

The method given by Liebig is as follows : 4 parts of basic verdigris and 3 parts of arsenious acid are separately dissolved in boiling vinegar; the liquids are mixed and evaporated until a precipitate of emerald green separates.

According to Ehrmann and Kastner, 70 parts of verdigris and 56 to 63 parts of arsenious acid are separately dissolved, the boiling solutions mixed and boiled until the dirty green precipitate changes into a bright green. In consequence of the rapid formation of the precipitate due to the boiling, the colour is pale green; if a dark-coloured product is required, the solutions are allowed to cool before mixing. After several days a very dark-coloured precipitate separates from the liquid; it consists of comparatively very large crystals of emerald green.

Manufacture of Emerald Green from Copper Sulphate.— There are two processes. Either the solution of copper sulphate is mixed with an acetate, by which it is converted into copper acetate, and then with arsenious acid, or copper arsenite is precipitated from the solution of copper sulphate and then converted into emerald green by treatment with acetic acid. If acetic acid can be cheaply obtained, the latter process is to be recommended.

Braconnet gives the following directions: 3 parts of copper sulphate are dissolved in a little boiling water, the hot solution is mixed with a hot solution of 4 parts of arsenious acid and four parts of potassium carbonate; the precipitate is then treated with 3 parts of pyroligneous acid. The dull-coloured precipitate at first formed is changed by the acetic acid to a bright green. The colour is more quickly produced when the liquid is heated nearly to boiling for several hours. When the desired shade appears the boiling is stopped and the precipitate at once separated from the liquid, by long contact with which it would give up arsenious acid and thus lose in shade.

According to Fuchs, 5 kilogrammes of lime and 25 kilogrammes of copper sulphate are dissolved in acetic acid, then a boiling solution of 25 kilogrammes of arsenious acid is added. The precipitate is at once formed; it is immediately separated from the liquid, washed, and dried; the top liquor, with the addition of arsenious acid, can be again used to precipitate the copper lime solution.

Emerald green has an extremely bright colour which surpasses other green mineral pigments in beauty. The larger are the microscopic crystals of which it consists the deeper is the colour. In order to obtain the desired deep shade of emerald green it is necessary to use very dilute solutions; from strong solutions the precipitate is instantaneously produced, in which case it is impossible to obtain large crystals. The covering power of emerald green, in consequence of its crystalline nature, is less the deeper the shade. It cannot be regarded as a permanent colour: very dilute alkalis and acids attack it. By the action of sulphuretted hydrogen it is discoloured, owing to the formation of black copper sulphide. It ought not to be used upon

...e, which withdraws acetic acid and produces yellowish-green copper arsenite.

This pigment should not be used for distempering or for colouring wall papers. By the action of the atmosphere on the arsenic it contains, arseniuretted hydrogen is formed—an extremely poisonous gas, very small quantities of which are sufficient to cause serious illness in the case of susceptible people. In spite of its poisonous nature, emerald green is still largely used. All possible shades of green can be produced from it, yellowish green or pale green, by admixtures of yellow pigments, such as chrome yellow, or white pigments, such as barytes or white lead. Many qualities of green known as palette green, Basle green, etc., are generally composed of such mixtures.

Mitis Green or Vienna Green is obtained by dissolving verdigris in acetic acid and adding a boiling solution of white arsenic in water: acetic acid is then added until the precipitate dissolves. On boiling the clear solution a green precipitate forms, which, when dry, is a deep bluish-green of characteristic shade. This pigment is now practically disused because of its poisonous nature, which is equal to that of emerald green.

Copper Stannate.—This green pigment, also known as Gentele's green, can be made by two methods. Gentele gave the following process: 59 parts of tin are converted into stannic chloride by solution in *aqua regia*; a solution of 131 parts of copper sulphate is added and copper stannate is precipitated from the mixture by caustic soda, on washing and igniting it acquires a pretty green colour. It may also be made by fusing 59 parts of tin with saltpetre; the potassium stannate is then dissolved in dilute caustic soda and copper sulphate solution added; the precipitate is washed and dried.

Copper stannate is tolerably durable, in this respect it

considerably surpasses emerald green. Only sulphuretted hydrogen has any considerable action upon it, turning it to a dirty brownish-green hue.

Kuhlmann's Green is a basic chloride of copper, obtained by heating 2 equivalents of lime with a solution of 3 equivalents of copper chloride. It is important that the copper salt should be present in excess. In shade Kuhlmann's green is very similar to emerald green, with which it agrees in retaining its colour by artificial light. It is a somewhat less pure green than emerald green; the difference is only perceived when the two pigments are directly compared. When, however, it is considered that it is far more permanent than emerald green and cheaper, a more extended use is indicated than is yet the case.

Elsner's Green is a species of lake, made by mixing a solution of copper sulphate with a decoction of fustic, adding a small quantity of stannous chloride and precipitating with caustic soda. To 100 parts of copper sulphate 10 to 14 parts of stannous chloride are used. According as the copper salt or fustic extract predominates, the colour of the precipitate inclines to blue or yellow.

Elsner's green is also sold under the name of "non-poisonous green". This description is incorrect; the pigment is indeed free from arsenic, but is poisonous on account of the copper it contains.

Casselmann's Green approaches emerald green in brightness. It consists of a compound of copper sulphate with copper hydroxide and water, $CuSO_4.3Cu(OH)_2.\frac{1}{2}H_2O$. This pigment is obtained by mixing solutions of 4 equivalents of copper sulphate and 3 equivalents of sodium acetate at a certain temperature. The author's experiments have shown that the best results are obtained when the solutions are mixed at a temperature of about $100°$ C. With this object the solutions are placed in vessels standing in a pan of boiling water; when their temperatures have risen to about 100° C. they are quickly mixed, the mixture stirred, and the precipitate allowed to settle. When the precipitate is cautiously treated with very weak caustic soda solution a rather deeper colour is obtained. This treatment should not be continued too long, or the precipitate may acquire an ugly bluish shade.

Lime Green is a mixture of copper arsenite and calcium sulphate. It is thus a pigment which ought no longer to be used, on account of its poisonous properties. It is made by boiling milk of lime with excess of arsenious acid so long as the latter is dissolved. To this solution of calcium arsenite, copper sulphate solution is added so long as a precipitate is formed. A mixture of copper arsenite and calcium sulphate is precipitated.

Patent Green is similar in composition to lime green. A solution of calcium acetate is made by adding quicklime or pure powdered limestone to acetic acid; copper sulphate solution is then added, when gypsum is precipitated and copper acetate remains in solution. A hot solution of arsenious acid is next added, and the precipitated copper salt mixed with the gypsum at the bottom of the vessel.

Copper Borate.—A pale green precipitate of copper borate is obtained by adding a solution of 3 parts of borax to a solution of 2 parts of copper sulphate. It must be dried at a very moderate temperature, or it will decompose. When perfectly dry the precipitate can be heated to a red heat without decomposition; according to the temperature employed different shades are obtained. It is most convenient to take tests out of the crucible from time to time during the ignition, and to quickly cool the crucible when the desired shade is obtained. When levigated this pigment may be used in oil, or as a porcelain colour.

Copper Silicate (Egyptian Blue).—When a solution of

water-glass is added to a solution of copper sulphate a pale green precipitate of copper silicate is obtained, which can be heated to redness without alteration. When 70 parts of white quartz sand, 15 parts of copper oxide, 25 parts of chalk and 6 parts of soda are fused together a glass is produced which, after pouring into water, grinding and levigating, exhibits a pretty blue and very permanent colour. It appears from the examination of the colours of the Egyptian mural paintings that this pigment was already known to the ancient Egyptians.

CHAPTER XXX.

VERDIGRIS.

VERDIGRIS is little used as a pigment, but is important to the colour maker, because copper colours are made from it. In wine-growing countries it is made from the marc at small expense and without much labour.

In commerce several kinds of verdigris are distinguished; they differ in physical and chemical properties. All verdigris consists of copper acetate, either alone or combined with varying quantities of copper hydroxide.

Blue Verdigris is made in large quantities in France and is commonly known as French verdigris (*vert de gris naturel*). It has the following formula:—

$$Cu(OH)(C_2H_3O_2) \cdot 2\tfrac{1}{2}H_2O.$$

The process adopted in southern France, especially in the neighbourhood of Montpellier, is as follows: copper plates are placed in layers in heaps of the freshly pressed grape residues, which always contain a certain quantity of must even when the most powerful presses are used. The process is conducted either in large heaps or in pots; in the first case care must be taken that the heaps are in a room whose temperature does not lie below 15° C. It is important to maintain this temperature, because the formation of acetic acid takes place with sufficient rapidity only at temperatures not below 12° to 15° C. Too high a rise of temperature in the heaps must also be avoided or a considerable quantity of acetic acid will be volatilised. The heaps should therefore not be made too large, otherwise the temperature cannot be kept within the proper limits.

The residues contain a considerable quantity of sugar, which can be transformed by fermentation into alcohol. If air has free access, the alcohol is at once oxidised to acetic acid, which is recognised by the acid smell given off. In order to make possible the entry of air into the interior of the heaps, rectangular wooden bars are introduced in piling up the heaps. These are carefully withdrawn when the heap is finished, so that it is traversed by passages through which air can enter. By the simultaneous action of air and acetic acid copper acetate is formed, and since copper is always in excess, the acetate produced is a basic salt. In consequence of the chemical processes, the temperature of the heap rises to 35° to 40° C.; it is endeavoured to attain this temperature by artificial heat. It has already been stated that too great a rise of temperature is accompanied by a considerable evaporation of acetic acid. The process is then finished in a short time, but the loss of acetic acid is remarkable. When the temperature is so regulated that it varies between 25° and 30° C. the process takes a normal course—four to five days may be regarded as its duration.

The grape residues are effective on account of the sugar, which is transformed into alcohol and then into acetic acid. Residues which have been extracted with water after pressing should not be used in the manufacture of verdigris; they contain but little sugar and hence undergo a very feeble acetic fermentation. When such residues are used the verdigris is often accompanied by black spots of copper sulphide upon the plates; this is due to incipient decomposition of the residues.

As a practical test for the termination of the process a strip of copper is plunged into the heap and left for some hours. It should be covered by a uniform coating of verdigris; if it is covered with small drops that is a sign that the process is not completed and the heaps must be left some time longer. The course of the process can be simply and safely followed by means of a thermometer. This is placed in a perforated copper tube and plunged into the interior of the heap. A continuous rise of temperature indicates a steady increase of chemical action; the temperature of the room is regulated in accordance. When the temperature in the interior of the mass begins to fall the process is nearing the end. More heat is now supplied to assist the chemical action. The process is at an end when, in spite of external heat, the temperature of the heap decreases and approaches the temperature of the air of the room.

Verdigris can also be made by placing sheet copper and grape residues in layers in pots, which are deposited in a room of fairly uniform temperature, such as a cellar. This method has the advantage that the formation of the verdigris is finished in a shorter time, which is more than counterbalanced by the labour and expense of filling the pots.

The copper plates, when removed from the heaps or pots, are covered by a crust of thin needle-shaped crystals of a bright green colour. They are shaken to free them from the adhering grape skins and seeds, and then treated with weak vinegar, in which they are dipped; they are then left standing on edge for several days. This treatment with vinegar has the object of converting the neutral salt formed on the plates into basic salt. The dipping into vinegar and exposure to air are repeated six to eight times, when the originally pure green crust on the plates is gradually changed into bluish-green verdigris. These processes are continued until the plates are covered uniformly by a layer of verdigris about 5 centimetres thick. The crust is then scraped off by copper knives and stirred with water to a paste, which is pressed in leather bags in rectangular moulds. The lumps are then slowly dried in the air.

Plates which have been once used give a larger quantity of verdigris in a second operation, which is to be ascribed to the fact that they possess a larger surface than unused plates. New plates are made more susceptible to the further action of the acid by dipping in strong vinegar, by which they become covered with a layer of copper acetate.

The verdigris obtained by this method consists of crystalline scales of a pale bluish-green colour, which produce a pale blue powder. In crude verdigris grape seeds are often found, and occasionally stalks and pieces of metallic copper. These admixtures are a consequence of the method of manufacture and cannot be regarded as adulterants. Gypsum is to be regarded as an adulterant. Verdigris often contains basic copper carbonate.

Verdigris behaves in a peculiar manner with water. In contact with a small quantity it swells to a bluish-green mass, which becomes quite warm. Neutral copper acetate and the basic salts $2Cu(C_2H_3O_2)_2 . Cu(OH)_2 . 5H_2O$ and $Cu(C_2H_3O_2)_2 . 2CuO . 1\tfrac{1}{2}H_2O$ are formed, the latter of which is insoluble.

By the addition of a larger quantity of water the basic compounds are decomposed, neutral copper acetate dissolves, and a mixture remains of the compound $Cu(C_2H_3O_2)_2 . 2CuO . 1\tfrac{1}{2}H_2O$ and a brown basic acetate containing still less acetic acid. On account of this peculiar behaviour of verdigris towards water some care is required when it is used as a water colour. A very dilute solution is not green, but has an indefinite shade.

Distilled or Crystallised Verdigris consists of neutral copper acetate, $Cu(C_2H_3O_2)_2 . H_2O$. It can be made from blue verdigris by treating with the amount of acetic acid required to completely neutralise the copper oxide, or by decomposing copper sulphate with the acetate of a metal which forms an insoluble or difficultly soluble sulphate.

Crystallised verdigris is very simply made from ordinary verdigris; the latter, whilst still moist, but containing only a small quantity of water, or the acetic acid would be too largely diluted, is brought into a pan and strong acetic acid poured over. Strong pyroligneous acid may be used, its empyreumatic odour is without influence on the quality of the product. The pan is heated until the contents almost, but not quite, boil. They are frequently stirred so that the particles at the bottom are brought in contact with the acid. A dark green solution is formed; when its colour no longer increases in depth it is allowed to stand until suspended solids sink to the bottom. The clear solution is then drawn off and rapidly evaporated in shallow copper pans. It is important that the liquid should not contain an appreciable quantity of free acid. When a crystalline skin begins to form on the surface of the solution it is drawn off into the crystallising vessels. These are made of glazed earthenware; wooden rods are placed in them, on which the crystallisation takes place. In order that large crystals shall be obtained, it is necessary to maintain a regular temperature in the room in which the crystallising vessels are placed. It is heated at the commencement, and the temperature is allowed to sink a little towards the end of the operation. From twelve to fourteen days are required to obtain large crystals. The mass of verdigris crystals adhering to each rod weighs 2·5 to 3 kilogrammes. The residual liquid is a saturated solution of verdigris, and is used in the next operation. It is only necessary to heat it with the residue from the first operation and some acetic acid in order to obtain a crystallisable solution of verdigris.

The residue in the pan contains metallic copper, grape stems and seeds, and basic copper acetate. The copper is extracted by moistening with acetic acid and exposing to the air, when verdigris is formed, which is added to a later batch.

It is often advisable to make verdigris from a soluble copper salt; the process varies somewhat according to the salt employed. From copper sulphate verdigris is obtained by the action of calcium acetate. When solutions of these compounds are mixed, calcium sulphate separates as a white precipitate, whilst a solution of the easily soluble copper acetate is left. It is only necessary to mix the solutions in equivalent proportions, separate the liquid from the precipitated gypsum and evaporate. Gypsum is somewhat soluble in water, so that when the solution is evaporated a double salt separates out first. This consists of copper calcium acetate; it may be used as a pigment, but is of less value than verdigris. It is advisable to prevent the formation of this double salt, which is accomplished by adding copper sulphate to a slightly acid solution of calcium acetate until a precipitate is no longer produced. The solution of copper acetate is boiled for several hours to effect the separation of the dissolved gypsum; at the same time the iron contained as impurity in the copper sulphate is also precipitated. The purified solution of verdigris is concentrated and allowed to cool, when a further quantity of gypsum separates; the solution is then evaporated to crystallisation.

If barium salts and lead acetate are cheap, barium acetate can be obtained by double decomposition. When copper sulphate is added to the solution of barium acetate so long as a precipitate is formed, a solution of verdigris is produced, which only requires evaporation to crystallise. The precipitate is barium sulphate, and can be used as enamel white. It obstinately retains a quantity of copper, which, though small, is sufficient to give it a greenish tint. By several washings with acetic acid this copper is removed; the acid can be used to obtain fresh quantities of verdigris, so that none is lost, and the copper retained in the precipitate is regained. The precipitate, after treatment with acetic acid, only requires thorough washing with water to produce permanent white, satisfactory in every respect.

Ammonia may also be used in the manufacture of verdigris from copper sulphate. Strong ammonia is left in contact with copper sulphate in a covered vessel for several hours; gentle warming accelerates the formation of copper ammonium sulphate. The liquid has then a fine blue colour; it should not contain sufficient ammonia to be perceptible by the smell. It is then gently warmed and acetic acid gradually added. On continued heating small green crystals separate from the solution, which is boiled with continual additions of acetic acid in small quantities until small crystals appear on the surface, when it is allowed to cool and the crystals are strained off.

"Distilled" verdigris forms deep bluish green crystals, which effloresce slightly in the air; they dissolve in 5 parts of boiling water and in 13·4 parts at 20° C. When the solution is boiled for some time the tribasic acetate separates; the liquid becomes brown and acquires an acid reaction.

German Verdigris does not differ essentially from ordinary verdigris. It is a compound of the following basic acetates: $2Cu(C_2H_3O_2)_2.CuO$ and $Cu(C_2H_3O_2)_2.2CuO$. The method is similar to that in which grape residues are used; where acetic acid is cheap the process is especially appropriate. In Sweden, where pyroligneous acid is made in large quantities and copper is also cheap, verdigris is made by arranging copper plates and flannel in alternate layers. The plates are frequently turned over and the flannel kept saturated with acetic acid. The same chemical reactions occur as in the process in which grape residues are used. When a layer of verdigris has formed on the copper plates they are taken apart, exposed to the air, and frequently wetted. The formation of the verdigris is thus completed. The difference between this process and that adopted in wine-producing countries lies simply in the use of pure acetic acid; the product is free from the mechanical impurities introduced by the grape residues. In properties German or Swedish verdigris is completely identical with blue verdigris.

CHAPTER XXXI.

CHROMIUM OXIDE.

Chrome Green.—Although very different substances are commercially known as chrome green, this name is usually applied to the very valuable pigment which consists of pure chromium oxide. There is hardly another pigment for the preparation of which such varied directions have been given and whose shade varies so much according to the manner of production.

The cheapest process is to ignite potassium bichromate with sulphur, extract the mass with very dilute sulphuric acid and wash the residue. The chromic acid is reduced by the sulphur; when the mass is treated with sulphuric acid sulphur dioxide is evolved, potassium sulphide and sulphate are dissolved, whilst pure chromium oxide remains. The larger the quantity of sulphur used the paler is the chromium oxide. The purity of the potassium bichromate has considerable influence on the shade of the pigment; if iron is present in any quantity a discoloured product always results. Convenient proportions are : 19 parts of potassium bichromate and 4 parts of sulphur, which produce $9\cdot 33$ parts of chromium oxide. If potassium bichromate free from iron cannot be obtained, the shade of the product is somewhat improved by treatment with dilute hydrochloric acid, in which ferric oxide dissolves more easily than chromium oxide. When the latter has been strongly ignited it is soluble with very great difficulty.

There are many formulæ for the preparation of chromium oxide by means of sulphur: in all the above statement holds good, that the more sulphur the paler the product.

According to A. Casali, a chrome green which satisfies every requirement is obtained by strongly igniting 1 part of potassium bichromate with 3 parts of burnt gypsum. After ignition the mass is boiled with very dilute hydrochloric acid. In this process the chromic oxide is produced according to the following equation:—

$$2K_2Cr_2O_7 + CaSO_4 = 2Cr_2O_3 + 2K_2SO_4 + 2CaO + 3O_2.$$

On boiling with hydrochloric acid the lime is dissolved; when the liquid remains distinctly acid after long boiling it is poured off and the chromium oxide washed with hot water and dried.

In another method ammonium chromate is very gradually heated. At a certain temperature the salt suddenly becomes incandescent and is converted into a dark green, almost black, mass which has a very similar appearance to dry tea-leaves. The green obtained after washing and powdering is handsomer the lower the temperature of the decomposition.

Chromium oxide may also be made in the wet way, but the product does not compare in beauty with that obtained in the dry way. When soda solution is added to a solution of chrome alum, a greyish green precipitate of chromium hydroxide is formed; on ignition of the washed precipitate pure chromium oxide remains.

In a similar manner chromium oxide is produced by precipitating with soda solution the solution of chromium chloride obtained by adding hydrochloric acid to a solution of potassium bichromate, and then alcohol in small quantities, so long as a reaction occurs and the green liquid becomes deeper in colour.

A handsome and bright chrome green is only obtained when the potassium bichromate is quite free from iron, a very small quantity of which has a very harmful effect upon the brightness of the pigment. The author has found that there is no great difficulty in rendering commercial potassium bichromate tolerably free from iron by recrystallisation. As much of the salt is dissolved in boiling water as it will take up, the boiling solution is quickly filtered and rapidly cooled with continual stirring; the fine crystalline meal is left on a strainer until no more liquid filters through, and then washed with a little cold water to remove the mother liquor. The salt obtained by this simple operation is of great purity and always produces chrome green of a good shade.

When chrome green contains ferric oxide it shows a blackish colour. From such chrome green, which has little value, a product of more pure colour can be obtained with a little care. By treatment with dilute hydrochloric acid the ferric oxide dissolves, even when it has been somewhat strongly ignited; chromium oxide loses almost all solubility in dilute hydrochloric acid after heating at a relatively low temperature. To facilitate the solution of the ferric oxide, the green to be treated is finely powdered; it is then covered with a mixture of equal parts of hydrochloric acid and water. After several days the mass is strained and washed with pure water until the acid reaction disappears. Comparative analyses have shown that almost the whole of the ferric oxide can be removed from chrome green in this manner; the brightness of the colour is considerably increased. It is, however, advisable to employ materials free from iron in the manufacture of chrome green.

Chromium oxide can be obtained as a colour of particular brightness, according to the process of Leune, by precipitating it very slowly from a solution in which it is contained in the green modification. For this purpose a solution of chrome alum is boiled until the violet colour has changed to green; the green modification of chromium oxide is now contained in the solution. It is cooled to about 10° C., and either freshly precipitated alumina or zinc carbonate added, very gradually and in small quantities. A fine green shade of chromium oxide separates; after washing and drying it is obtained in a condition in which it can be used as a pigment.

Chromium oxide can indeed be obtained by this process, but it is not distinguished by particular beauty; the author has never been able to obtain, even by using the precipitant in very small quantities, a chromium oxide surpassing in shade that obtained by the direct precipitation of a chromium salt with an alkali.

The emerald green described in the following chapter is obtained in a similar manner by precipitation with zinc oxide.

CHAPTER XXXII.

OTHER GREEN CHROMIUM PIGMENTS.

Guignet's Green is a chromium oxide made in the dry way; it[1] is obtained by grinding 1 part of potassium bichromate with 3 parts of pure boric acid and water, drying and heating to a dark red heat with access of air. The hot mass is brought into water; by boiling repeatedly with water it is freed from the boric acid which it obstinately retains. In order to remove the last portions of boric acid it is necessary to boil with sulphuric acid and then with caustic soda; this treatment may be omitted for technical purposes, since the small quantities of boric acid contained in the pigment are not harmful.

Guignet's green is distinguished by the great resistance it offers to chemical reagents; it is largely used in painting and calico printing. Pale shades are obtained by additions of barytes.

Emerald Green.—This pigment must not be confounded with the more important copper compound previously described, to which the name emerald green is usually applied. It consists of chromium hydroxide obtained by precipitating a solution of the green modification of a chromium salt by zinc hydroxide. After careful washing it has a dull green shade; it is a very durable pigment.

Chrome Green Lake is a mixture of chromium oxide with alumina; it is obtained by precipitating a solution of alum and a chromium salt by a soda solution. The precipitate, which contains aluminium and chromium hydroxides, on ignition takes a paler shade in proportion to the quantity of alumina it contains. For the chromium salt a solution of potassium bichromate, which has been allowed to stand with sulphuric acid and alcohol until it has acquired a pure green colour, may be used.

Turkish Green.—This pigment, which retains its fine green colour in artificial light, is prepared in a peculiar manner: 40 parts of alumina free from iron, 30 parts of cobalt carbonate and 20 parts of chromium oxide are thoroughly ground in a mortar. The mixture is placed in a porcelain tube, which is exposed to a strong white heat whilst pure oxygen is led through. The oxygen may be replaced by air if it is previously heated and under pressure.

Another recipe for Turkish green is to intimately mix 4 parts of freshly precipitated alumina with 3 parts of cobalt carbonate and 2 parts of chromium oxide, heat the mixture in a crucible at a white heat, powder and levigate. Turkish green has a characteristic bluish-green colour; the shade is inclined to blue or green by increasing the quantity of cobalt carbonate or chromium oxide.

Leaf Green is a pale green, very durable pigment similar to chrome green lake. It is obtained by igniting mixtures of chromium oxide with pure aluminium hydrate; the paleness of the shade is in proportion to the amount of alumina.

Chromium Phosphate Pigments.—Several chromium phosphates are used as pigments; the more important are described here.

Arnaudan's Green is chromium metaphosphate; it is obtained by intimately mixing 128 parts of neutral ammonium phosphate with 149 parts of potassium bichromate by long grinding, and carefully heating the mixture at 170° to 180° C., but not higher, until the mass is pure green. It is then brought into hot water and thoroughly washed; it has a very handsome green colour, which is not altered in artificial light.

Ammonium arsenate may be used in the place of ammonium phosphate; a handsomer green is then obtained, which is, however, very poisonous.

Plessy's Green is essentially a chromium phosphate mixed with variable amounts of chromium oxide and calcium phosphate. It is obtained by boiling a solution of 1 part of potassium bichromate in 10 parts of water with 3 parts of a solution of acid calcium phosphate and 1 part of sugar, until the whole has become deep green. The chromic acid is reduced by the sugar, so that the precipitate contains chromium phosphate, chromium oxide and neutral calcium phosphate.

These two chromium phosphate greens are very stable towards chemical reagents, and very durable towards atmospheric influences.

Schnitzer's Green.—Thirty-six parts of sodium phosphate are melted in its water of crystallisation, 15 parts of potassium bichromate and 14 parts of Rochelle salt are then added. A large dish must be used for the fusion, since the mass effervesces. The colour changes gradually from yellow to green. When it is pure green the heating is stopped, and, after cooling somewhat, as much hydrochloric acid is added as the mass will absorb; then, after standing some time, it is washed with cold and finally with boiling water. On account of their great durability, the chromium phosphate pigments are well adapted for paperhangings, calico printing and oil painting.

Chromaventurine is a glass coloured green by chromium oxide. It is hardly used as a painters' colour in the ordinary sense of the term; it is applied in porcelain painting as an under-glaze colour, and is largely used for colouring glass.

Chromaventurine is most simply made by the process of Pelouze: 250 parts of quartz sand, 100 parts of soda, 50 parts of calcium carbonate and 40 parts of potassium bichromate are fused together. The presence of iron would have a very harmful effect on the shade of the product. Quartz sand generally contains a small quantity of ferric oxide, which should be extracted by treatment with strong hydrochloric acid when a product of the best colour is required. The calcium carbonate should also, as far as possible, be free from ferric oxide.

Chrome Blue (Garnier).—A mixture of 48·62 parts of potassium chromate with 65 parts of fluorspar and 157 parts of silica is fused in a crucible lined with coal-dust.

CHAPTER XXXIII.

GREEN COBALT PIGMENTS.

FROM mixtures of chromium oxide, cobalt carbonate and alumina in different proportions a series of pigments can be obtained varying in shade between pale blue and bluish green. On account of the great stability of these colours at the highest temperatures they are of great importance in porcelain painting, for which they are principally used. It is necessary that the porcelain should be free from iron, since ferric oxide forms a black compound with chromium oxide, very small quantities of which would be sufficient to considerably damage the fineness of the colour.

Cobalt Green, which is also known as Rinmann's green or zinc green, is a compound of the oxides of cobalt and zinc. It is always produced when a cobalt compound is ignited with zinc oxide. Rinmann's green has not so deep a colour as the poisonous emerald green, but it is distinguished by extraordinary durability, and deserves to be more largely used than it is at present. This pigment is most simply made by precipitating the mixed solutions of a zinc salt and a cobalt salt. It is also formed by moistening pure zinc oxide with a cobalt solution and igniting.

In precipitating the mixed solutions of zinc and cobalt salts products of different shades are obtained, according to the proportions in which the salts are used. If equivalent quantities are used, an almost black product is obtained, quite useless as an artists' colour. The best result is obtained by mixing intimately pure precipitated cobalt carbonate with zinc oxide and igniting the mixture; a mixture of 9 to 10 parts of zinc oxide with 1 to 1·5 part of cobalt carbonate gives colours between pale green and dark green.

Especially fine Rinmann's green is obtained by igniting cobalt arsenate with zinc oxide and arsenious acid. The addition of the last-named substance can only have the object of preventing the temperature from rising too high, which might injure the beauty of the colour; arsenious acid is volatile at a fairly low temperature.

Cobalt green is also obtained by evaporating a solution of cobalt nitrate and zinc nitrate and igniting the residue, or by igniting a mixture of the sulphates. In the latter case a tolerably high temperature is necessary in order that the sulphates may be decomposed.

The author has found that a particularly handsome product is obtained by mixing pure zinc oxide with a dilute solution of cobalt chloride, drying and then slowly heating to redness in a crucible with a well-fitting lid. Towards the end of the operation the heat is considerably increased and maintainèd for a short time, after which the mass is rapidly cooled.

CHAPTER XXXIV

GREEN MINERAL PIGMENTS

Manganese Green. Rosenstiehl's Green.—This pigment, which is handsome but difficult to produce, consists of barium manganate. It can be obtained by several methods, of which produce a good green, but with very unequal properties. Manganese green made from barium nitrate has little durability, when made from barium hydrate it is more stable but very fugitive.

Barium manganate is most easily obtained by precipitating a boiling solution of potassium manganate by barium chloride. The almost white precipitate becomes nearly white when washed and dried, but when it is gradually heated to a moderate heat in a dark red heat it acquires a fine green color. The heating must be carefully conducted, if the temperature rises too high the color changes to a dirty greenish grey, owing to the reduction of the manganic acid.

The pigment is also made by heating 14 parts of manganic oxide, 4 parts of barium nitrate and 4 parts of baryta with access of air until the desired shade appears. The mass is then ground in a continuous current of water until it is changed into a very fine powder and nothing more is carried off by the water. The best result is obtained by Rosenstiehl's process: 4 parts of barium hydrate, 2 parts of finely-powdered barium nitrate and 0.5 part of artificial manganese dioxide are being intimately mixed is moistened and heated to a dark red heat. The fused mass is boiled out with water and the residue dried under a bell jar, under which are dishes of sulphuric acid and caustic potash, the former of which absorbs the water, whilst the latter keeps the air free from carbonic acid, which would injure the shade of the moist substance.

Manganese green is amongst the pigments more recently discovered. On account of its high price it has found but limited use up to the present. It may be used in any vehicle, and is distinguished by great permanence.

In regard to the preparation of this colour by Rosenstiehl's method, which of all gives the best results, it should be observed that the shade largely depends on the quantity of barium hydrate used; the greater this is the more the shade inclines to blue. If it is desired to produce a colour of a greener tone than is produced by the fusion, this can be accomplished by boiling for a very long time with very weak hydrochloric acid, which extracts a portion of the base from the compound and produces a deeper shade.

Böttger's Barium Green.—A beautiful green is obtained by the process given by Böttger, as the author has found. It consists of barium manganate. The process is, however, somewhat costly, so that the pigment would only be available for artists' purposes.

A solution of potassium manganate is first made by gradually adding 2 parts of very pure pyrolusite, finely powdered, to a fused mixture of 2 parts of caustic potash and 1 part of potassium chlorate, bringing the mass to a low red heat after the introduction of the whole of the pyrolusite, and finally extracting with water, in which the potassium manganate dissolves to a fine emerald green solution. This process is unattended by difficulty if the pyrolusite contains a sufficient quantity of manganese peroxide and is in a sufficiently fine powder, which is most important. Potassium manganate is a very unstable substance. Cold water must be used to dissolve it, and the solution should not long stand exposed to the air, but should at once be used to obtain barium manganate.

When the solution of potassium manganate is mixed with a solution of a barium salt, a handsome violet precipitate immediately forms. This is washed with water and quickly ground with ¾ to 1 part of its weight of barium hydrate. The mixture is heated in a copper dish with constant stirring to a low red heat; the colour changes gradually to a very fine green. When the mass has reached the proper shade it is treated with cold water to remove excess of barium hydrate, until the washings show no trace of alkaline reaction.

Manganous Oxide is occasionally used as a green pigment, especially for painting metal work; it is obtained in the following manner: Manganese sulphate solution is precipitated by soda solution, and the manganous carbonate obtained strongly ignited in a crucible. In order to avoid oxidation of the manganous oxide, which would readily occur in the process, it is necessary to prevent air from reaching the substance. With this object the crucible in which the heating is conducted is covered by another from which the bottom has been removed and which is filled with coal. The air entering the crucible on cooling must pass through the layer of glowing coals, by which it is deprived of oxygen.

Manganese Blue is obtained, according to G. Bong, by igniting a mixture of 3 parts of quartz, 6 parts of soda ash, 5 parts of limestone and 3 parts of manganese oxide, or of 3 parts of quartz, 8 parts of barium nitrate and 3 parts of manganese oxide, air being admitted and reducing gases excluded. All materials must be free from iron. The quantity of manganese oxide regulates the depth of the colour, but not its shade; by increasing the soda a greener, by increasing the quartz a more violet, shade is obtained.

CHAPTER XXXV.

COMPOUNDED GREEN PIGMENTS.

MIXTURES of a yellow and a blue pigment produce a green; according as one or the other predominates colours are obtained inclining to yellow or blue.

In some cases the mixing can be accomplished in the actual production of the pigments, so that a green precipitate is directly produced. This is, however, rarely the case; generally the compound pigments are obtained by simply mixing the two colours. The mixing can be done either in the dry or the wet way; colours in the wet state are more easily mixed than when dry, so that even when dry colours are employed the mixing is done with the addition of water. Although the mixture then requires a second drying, this method is still to be recommended, in the first place, because the mixing can be more quickly done in consequence of the greater mobility of the mass, and in the second, because the formation of poisonous dust is completely avoided.

The mixing is generally accomplished by mechanical arrangements. If dry colours are to be mixed, rotating cylinders may be used; these are filled with the materials to be mixed, well closed and rotated about the axis as long as may be necessary. It should be observed that when colours which have a very different specific gravity, such as chrome yellow and Prussian blue, are to be mixed, the cylinders must be rotated for a much longer time than when colours of approximately equal specific gravity are to be united.

In working in the wet way, which is to be preferred, sufficient water is added to the colours to make a paste thin enough to be mixed by spatulas; the mixture is thoroughly stirred and the pulp ground in ordinary mills until the mixture is quite uniform. Already whilst on the mills the pulp becomes thicker in consequence of evaporation; when the grinding is finished it is spread out in thin layers so that it may dry as quickly as possible. This is important in the case of mixtures whose constituents have very different specific gravities, otherwise the heavier pigment may sink to the bottom of the paste and the mass thus lose its uniformity.

The compound green pigments come into the market under most varied names, which are often entirely without connection with their chemical composition. Such names are: mineral green, English green, oil green, green vermilion; the most common are chrome green and Brunswick green. It should be noted that these names are more strictly applied to simple colours previously described.

Chrome Green is generally made by mixing deep chrome yellow and Prussian blue; it can naturally be obtained in all possible shades. The brightness of this already handsome pigment can be considerably increased by the addition of a small quantity of indigo carmine. In this case the mixing is best done by adding a solution of the indigo carmine to the stiff paste and again sending the mixture through the mill.

By additions of white substances paler shades of chrome green are obtained; levigated terra alba or white pipe-clay can be advantageously employed.

Elsner's Chrome Green is obtained by preparing a solution of yellow prussiate of potash and potassium chromate and another solution of lead acetate and ferric chloride. The two liquids are mixed with vigorous stirring; according to the proportions of the materials a blue or a yellow shade of green is produced.

Silk Green.—Forty-one parts of lead nitrate are dissolved in 20 to 30 times the weight of water, the solution is boiled in a copper pan, and, according to the shade required, 10 to 30 parts of fine Chinese blue are added. After well stirring, a solution of 10 parts of potassium bichromate and 1 part of nitric acid is poured into the boiling liquid, the mixture is again well stirred, the precipitate allowed to settle, washed and dried. The green pigment obtained in this way has a peculiar silky lustre, hence the name "silk green".

Natural Green is a mixture of Guignet's green with picric acid; it is used in making artificial flowers instead of emerald green.

Non-arsenical Green was proposed as a substitute for emerald green, which, however, it does not equal in brilliance. It is obtained by mixing copper blue (basic copper carbonate) with chrome yellow, chalk and ferric oxide, in somewhat variable proportions. It usually contains 80 to 82 per cent. of the copper compound and 13 to 15 per cent. of chrome yellow.

In making mixtures of pigments it is to be remembered that chemical actions may occur in the mixture; colours which act upon one another should not be mixed: the mixture would contain the cause of its own destruction. For example, a lead pigment should not be mixed with one containing sulphur in the form of a sulphide or sulphate. By following this important rule the number of pigments which may be mixed together is considerably diminished; the advantage is that durable colours are produced, which do not change in a short time to an unrecognisable shade.

CHAPTER XXXVI.

VIOLET MINERAL PIGMENTS.

Chromic Chloride.—The violet modification of chromic chloride is a compound which is little used, but deserves the highest attention on account of its durability and beauty of shade. Up to the present chromic chloride has been almost exclusively used for colouring paper hangings under the name of chrome bronze, which it has received in consequence of its property of imparting a peculiar metallic lustre to paper upon which it is rubbed. It is also possible to fix this substance upon fabrics, which thus acquire a similar metallic lustre.

Pure chromic chloride forms beautiful peach-blossom coloured scales, which can only be obtained by treating chromium oxide in a certain manner with chlorine. The pure substance is practically insoluble in water, but if the water contains but a trace of chromous chloride the chromic chloride readily dissolves to a green solution. In the preparation of this compound it is necessary to avoid the least trace of chromous chloride, otherwise the product will begin to change when it comes into contact with moist air.

Pure chromic chloride is obtained by proceeding according to the method first given by Wöhler: the simple apparatus necessary is depicted in Fig. 29. Pure chromic oxide is first made by any suitable process; this is made into a paste with charcoal, starch paste and water, and formed into small balls which are heated in a crucible to a white heat. The residual intimate mixture of chromic oxide and carbon is loosely placed in a crucible which stands in a furnace; a porcelain tube is cemented into the bottom of the crucible, it passes through the ashpit to connect with a chlorine apparatus. Upon the crucible another larger one is placed, which has a small opening in the bottom and serves as a receiver for the sublimed chromic chloride. The heating and current of chlorine are started at the same time, the fire is regulated so that the lower crucible glows brightly. This

Fig. 29.

is continued for an hour, during which a rapid current of chlorine must pass into the crucible; the fire is then removed and a slow current of chlorine passed through the apparatus until it is quite cold.

If the operation has succeeded the lower crucible is found empty and the interior of the upper covered with the magnificent peach-blossom crystals of chromic chloride. A small quantity of the chromic chloride is tested as to its behaviour towards water; if it remains unaltered the whole may be washed, but if it dissolves to a green solution it must be again ignited in a current of chlorine. The washing is necessary to remove small quantities of aluminium chloride formed by the action of chlorine on the clay of the crucible.

Manganese Violet, or Nuremberg Violet, consists of manganic phosphate. It is obtained by fusing glacial phosphoric acid with pure pyrolusite, boiling the melt with ammonium carbonate, filtering the solution, evaporating to dryness, and again fusing the residue. After boiling with water this forms a handsome violet powder; it is suitable for use as a very durable artists' colour.

Shades of this pigment inclining to blue are obtained when a ferric compound is added in the first fusion; the more iron ore is used the more intense is the blue tone.

Tin Violet or Mineral Lake is obtained as a violet mass by igniting an intimate mixture of 100 parts of tin dioxide with 2 parts of chromic oxide. This pigment is completely permanent in air; it is suitable for printing paper hangings and for colouring faience.

Copper Violet, Guyard's Violet.—According to J. Depierre, this pigment is obtained by precipitating a solution of copper ammonium sulphate with potassium ferrocyanide, washing and drying the precipitate and heating in a porcelain dish. At 170° C. cyanogen and ammonia are evolved, the mass takes up oxygen and becomes violet; it then resists the action of dilute acids and alkalis, and as a pigment has great covering power. On heating to 200° C. it turns blue, and at 240° to 250° C. becomes greenish.

CHAPTER XXXVII.

BROWN MINERAL PIGMENTS.

Lead Brown.—When red lead is treated with nitric acid, lead monoxide is dissolved, whilst deep brown lead peroxide remains. When the action of the nitric acid is finished the residue is well washed and dried. Lead peroxide is now extensively used for lucifer matches; it readily gives up oxygen on heating, and thus accelerates the ignition of the composition. Brown match-heads owe their colour to lead peroxide.

Manganese Brown consists of manganic oxide; it occurs in nature, but seldom in a state of sufficient purity to be used as a pigment; it is, therefore, generally made artificially. The process is very simple. A solution of manganese sulphate is precipitated by caustic soda solution, the manganous hydroxide produced is quickly changed by the action of the air into manganic hydroxide. The colour is most rapidly developed when the precipitate is spread out in thin layers; it is well washed after it has changed to brown.

Pyrolusite Brown is manganese peroxide in a state of fine division. It can be made from the residual liquors of the preparation of chlorine by adding sodium hypochlorite. A brown precipitate results, which is kept in contact with the liquid until it no longer changes in colour. It is then washed with water slightly acidified with sulphuric acid, and finally with pure water. When finely powdered, manganese peroxide has a very fine brown colour. It is quite unaltered by the

PIGMENTS.

...its cheapness, deserves...

Prussian Brown.—When Prussian blue is heated in air it is converted into a brown mass... the amount of impurity in the Prussian... Prussian blue quite... and a brown of the same shade is... latter can be shaded by...

Iron Brown.—By igniting a mixture of 100 parts of finely... barium... with... parts of common salt... which is of a deep reddish... temperature... It is a cheap and durable...

Copper Brown.—Addition to different solutions of... sulphate and... Epsom salts are... sodium carbonate gradually... a brown mass... The precipitate is... filtered... brown is obtained...

...solutions of 4 parts of copper sulphate... with... 1 part of ferrous salts are mixed, potassium ferrocyanide solution added,... filtered. The redness of the shade... sulphate.

Hatchet Brown is a copper potassium ferrocyanide. It is obtained by gradually... potassium... copper sulphate... potassium ferrocyanide... different compounds according to the proportion of the copper salt in excess... somewhat larger...

Chrome Brown.—To a solution of potassium chromate... copper sulphate is added, a precipitate of the compound...

$CuCrO_4.2CuO.2H_2O$ is obtained. On drying it acquires a fine brown colour; it forms a very stable pigment.

Chrome brown can also be made by dissolving 10 parts of potassium bichromate in 20 parts of water, heating to boiling, adding 13·5 parts of solid copper chloride and then gradually a boiling solution of 10 parts of soda in 20 parts of water, until effervescence no longer takes place. On cooling, chrome brown separates as a soft brown precipitate.

Cobalt Brown is a very durable compound pigment of a pleasing shade. It is obtained in various hues by adding ferric oxide to the mixture of alumina and a cobalt salt used for making cobalt blue. The process may also be carried out by igniting ammonia alum with cobalt sulphate and ferrous sulphate. The temperature required in this case to obtain a bright colour is very high, and must be maintained for a long time to decompose the whole of the ferrous sulphate.

The pigment can be obtained at a far lower temperature when ferric chloride is used in place of ferrous sulphate. The mixture is made by grinding together 5 parts of cobalt hydroxide and 25 parts of ammonia alum, then adding a solution of ferric chloride, rapidly drying the whole to a powder, and whilst still hot filling it into the crucible in which the ignition is performed. When small quantities of ferric chloride are used chocolate brown shades inclining to violet are produced; the larger the quantity of the iron compound the more pure is the brown.

It has been stated that it is important, in the preparation of cobalt pigments, to prevent the entry of fire gases to the heated mixture; the reducing action of the gases would materially damage the beauty of the colour. It has been proposed to place a small quantity of mercuric oxide at the bottom of the crucible. This would decompose on heating, and create an atmosphere of pure oxygen in the crucible. Apart from the cost of the mercuric oxide, which is considerable, it would only be effective at the commencement of the operation, since it is completely decomposed at a low red heat. The author has found an addition of pyrolusite considerably more effective. This may be applied by spreading a small quantity, about 5 per cent. of the mixture to be heated, on the bottom of the crucible and covering it with powdered glass, upon which the colour mixture is placed. An alternative course is to place the crucible in a second, filling the space between the crucibles with powdered pyrolusite. At a strong red heat oxygen is slowly evolved from the pyrolusite; the entry of fire gases into the crucible is thus prevented. In using the first method the crucible is frequently spoilt, whilst in the second method the surrounding pyrolusite protects it, so that it can be used again.

CHAPTER XXXVIII.

BROWN DECOMPOSITION PRODUCTS.

Humins.—Wood decomposes, like many other substances of organic origin, to form compounds of a deep brown colour, which are known as humins, on account of their occurrence in the humus of tilled soil. The numerous compounds found in humus have all a deep brown colour, which, together with their great stability, makes them very suitable for use as pigments.

By various methods substances can be made so rich in humins that they can be used as pigments. For this purpose sugar, starch, young plant fibres and beet-sugar molasses may be used, which are converted into humins with great rapidity when they are heated with water. In this way the author has obtained handsome colours from sawdust. These humins are most easily made by the following process: Thick beet-sugar molasses are cautiously heated with 5 per cent. of caustic soda in a very capacious iron pan. The already dark mass soon becomes quite black when seen in thick layers, and evolves a considerable quantity of gas; if the heating is too rapid the soft mass may boil over out of a very large vessel. When the evolution of gas diminishes the heat is increased and the mass frequently stirred. With some practice it is possible to tell when the action is finished from the smell, which is at first sweetish but later of a characteristic nature. At first tests should be repeatedly taken from the mass; these are considerably diluted with water until the liquid begins to be transparent. When two tests taken at an interval show no difference of colour the heating is stopped. The whole mass is then poured into water in order to dilute the alkali, so that it will not destroy the strainer; the soft mass, which in the wet state appears quite black, is washed with water until the washings are neutral. The humin brown made in this way, when ground with oil or gum solution, produces a very handsome brown of great covering power and warmth of shade, also distinguished by complete indifference towards chemical reagents.

The pigments we have designated humins contain a very large quantity of carbon, to which they owe their dark colour. When peat or lignite is treated with caustic soda in a similar manner, good shades of brown are obtained; they are, however, surpassed in beauty by the brown from molasses. A handsome but costly brown is obtained when crude spirits of wine are heated with fuming sulphuric acid. Equal volumes of alcohol and sulphuric acid are used; the mixture is heated in a retort connected to a condenser, which is required on account of the combustible vapours evolved from the mixture. When the mass is quite black the heat is withdrawn; the residue is diluted with water, and soda solution added so long as effervescence occurs. On filtering a very soft brown powder is left, which forms a very handsome and durable pigment.

Bistre.—When soot is produced at a very low temperature it is very lustrous, and, in addition to carbon, contains a notable quantity of the products of dry distillation. When this soot is powdered and treated with water the latter substances are dissolved; boiling water should be used. When water is no longer coloured the soot is suspended in a large quantity of water and subjected to a process of levigation. The fine powder obtained by repeated levigation has, when dry, an ugly brown colour, but when ground it acquires an extremely warm shade.

CHAPTER XXXIX.

BLACK PIGMENTS.

CARBON occurs in nature in many different forms—the diamond, graphite (black lead), and purified lamp black are, from the chemical point of view, one and the same substance, namely, carbon. In colour making the non-crystalline form of carbon, which is pure black, is alone used; almost all black pigments used in painting are composed of tolerably pure carbon. In whatever way and from whatever materials carbon is made for use as a pigment, the manufacturer must always endeavour to obtain it as pure as possible, and in a condition of the finest division; the depth of colour and the covering power depend on these two conditions. Carbon only shows a pure black colour when it is pure; if it contains relatively small quantities of impurities, it has a more or less brown shade.

At first sight it appears to be a very simple matter to obtain black pigments from organic materials; it is simply necessary to expose them in the absence of air to a temperature high enough to decompose them, carbon is then left as a residue. In spite of the apparent simplicity of this operation there are many difficulties. The preparation of good black pigments demands considerable practice.

There are two methods by which carbon can be made for use as a pigment. One has been indicated, the heating of organic materials in the absence of air, *i.e.*, their dry distillation; the second, which produces the best qualities of black, consists in burning substances very rich in carbon with a very small supply of air. The product of this process, commonly known as soot, is very different in external appearance from carbon made by dry distillation. The latter forms an easily powdered mass, which only acquires lustre by grinding with oil, whilst soot consists generally of very light flocks, which, in consequence of their greasy nature, exhibit a peculiar velvety lustre.

In commerce there are many qualities of black, the names of which are generally chosen quite arbitrarily. The so-called ivory black is an example of this. At some period it was observed that ivory produced a very fine black when carbonised; this black was made for a long time from refuse ivory. When it was found that a similar black could be made quite as well from much cheaper materials, it was no longer made from ivory. The name, however, remains in trade to the present day: it has come to be regarded as a description of quality. By the name of ivory black a fine black pigment is understood; it is immaterial to the consumer whether it is made from ivory or not: to him the quality of the pigment, but not its source, is important.

According to the method of production, the black pigments can be divided into charcoal blacks and soot blacks.

Charcoal Blacks.

This term is applied to the pigments obtained by heating organic substances in the absence of air. The charcoal blacks have two valuable properties: they are easily made, and they exhibit a pure black shade which it is far more difficult to obtain with the soot blacks. It is very difficult to convert these pigments into the requisite state of fine division. They cannot be levigated on account of their low specific gravity, which causes them to settle very slowly in water. It only remains to convert the charcoal into a very soft powder by grinding, but even then the pigment has but little covering power, because it is not possible to destroy the organic structure of the substance carbonised. When the finest charcoal powder, made by carbonising any organic substance, is examined under a powerful microscope the structure of the particles is at once recognised, so that it is almost always possible to decide whether the charcoal is of plant or animal origin. A very careful examination may show what particular part of the plant has been carbonised. If a manufacturer succeeded in producing by direct carbonisation a black with the covering power of a soot black, that black would soon be the only one found in commerce.

The charcoal blacks come on the market under various names—ivory black, bone black, vine black, Frankfort black, Paris black, etc. The ivory blacks are the best quality, but recently true vine black, *i.e.*, made from grape residues, has come into use.

True Charcoal Black.—In charcoal-burning wood is burned in great piles with restricted air supply; the greater part of the carbon contained in the wood remains behind as charcoal, which is used for fuel. The charcoal made from hard wood, such as maple or beech, is not well suited for pigments; the lighter and more porous the wood, the more pure is generally the shade of the charcoal, and the more easily it can be ground. When, therefore, charcoal is to be used as a pigment it should be made from wood of a porous nature; the wood of the lime, black-alder and spindle tree is particularly adapted for this purpose. Spent tan bark is a very cheap, and at the same time suitable material for this purpose. It generally consists of oak bark, and has lost the greater part of the salts it contained by long contact with water; also the woody substance of the bark has suffered a change which is favourable to carbonisation. This is especially the case when the bark has been stored for some time; a considerable quantity of the humins previously mentioned has formed in it, and these are easily decomposed at a low temperature.

The charcoal, however obtained, is ground to a soft powder. When it has reached the proper degree of fineness it should be repeatedly washed with water to remove the salts. When instead of pure water a very dilute acid, such as hydrochloric, is used, practically all the salts dissolve, and the residue consists of nearly pure carbon.

Vine Black.—In wine-producing countries a good and cheap black can be made from the residues of the wine manufacture; this is the so-called vine black. Either the pressed grapes or the lees separated in the fermenting vessels may be used.

Vine Black from Wine Lees.—The lees always contain a considerable quantity of liquid, they must be thoroughly dried before they are carbonised; this is most simply done by spreading out the pasty mass in a thin layer, and exposing it to a temperature of about 100° to 120° C. In drying the volume considerably diminishes. The lees are changed into a brown mass which is easily powdered; it is packed into barrels while still warm, and can be kept for a considerable time without alteration. Fresh lees can be kept but a short time; they are quickly destroyed by a rapid fermentation.

The dried lees are carbonised in iron tubes, protected from the fire by a thin coating of clay mixed with chopped hair so that the coat more readily adheres to the iron. Old stove pipes or cast-iron gas or water pipes may be used. The tubes are about one metre long; they are closed with well fitting covers, in one of which is a small opening through which the gases produced on heating can escape. These covers are fastened on air-tight with clay. The process begins by affixing the unperforated lid, and packing with a wooden rammer the dry lees as tightly as possible into the tube; the second lid is then luted in place. The tubes are placed near one another in a suitable furnace and first slowly heated at the back, *i.e.*, the end closed by the unperforated cover. At the commencement of the operation heat must be gradually applied. With too large a fire the products of dry distillation might be evolved in sufficient quantity to force the covers from the tube; air would then enter and the contents be burnt. When the hinder part of the tube is red hot, the heating is conducted forward, and finally the whole length of the tube is brought to a good red heat. The termination of the operation is shown by the disappearance of the pointed flame of the products of distillation which protrudes from the opening in the cover of the tube. When this occurs, the fires are drawn; the tubes are left to cool in the furnace until they can be taken out by the hands protected by wet cloths. The covers are at once taken off and the hot contents emptied into a large tub filled with water, which is thus soon raised almost to boiling point. The time required for washing is thus considerably shortened. The charcoal falls from the tubes in lumps, which soon fall to a fine powder in the water. Wine lees consist of a mixture of yeast cells and small crystals of tartar (a mixture of the tartrates of potassium and calcium). On ignition these salts are converted into carbonates, by which the particles of carbon are united. Potassium carbonate is very readily soluble in water; in contact with the warm water it dissolves in a very short time, the carbon particles then fall apart.

When the charcoal has settled to the bottom of the vessel the liquid is drawn off. It can be utilised in a colour works, since it is a fairly strong solution of potassium carbonate. The charcoal is mixed with calcium carbonate, arising from the decomposition of the calcium tartrate, and with the other insoluble salts which yeast ashes contain in some quantity. These salts would prevent the proper grinding of the charcoal; they are therefore removed by treatment with hydrochloric acid after the water has been drawn off as completely as possible. A small quantity of hydrochloric acid, diluted with an equal volume of water, is poured over the charcoal, an effervescence of carbonic acid follows, the easily soluble calcium chloride is formed and the remaining salts are also dissolved.

The charcoal made in this way is very pure. After washing, grinding and drying, it forms a pigment whose shade leaves nothing to be desired. The drying must be conducted at a low temperature; charcoal in such a fine state of division is very easily inflammable.

Vine Black from Pressed Grapes.—The material contains the stems and pressed remains of the grapes after the must has been expressed. After-wine, spirits or vinegar can be obtained before it is used for vine black. The process is exactly the same as for wine lees. The charcoal taken from the tubes has rather more coherence than that made from lees and must be ground. The black is a very useful pigment, but is inferior to the black from lees. In regard to the latter, it should be stated that it is a particularly good black pigment, and when finely ground is as well suited for the preparation of the finest blacks, such as are used in copper-plate printing, as the far more costly soot black.

The black from grape residues, or from grapes themselves (the poor grapes removed in pruning the vine are used), can be employed with advantage for many purposes for which the best black is not necessary. The greater number of the pigments sold under the names of Frankfort and Paris black consist of this substance.

Bone Black or Ivory Black.—Bone black, for of ivory black only the name now exists, is distinguished by a peculiar quality which it owes to the structure of the raw material from which it is made. This raw material is animal bones, which largely consist of incombustible materials (bone ash);

these form a delicate framework whose interstices are filled by organic matter. When bones are carbonised, the carbon resulting from the decomposition of the organic matter is deposited on the incombustible framework of bone ash and thus acquires a very great surface.

Carbon is well known to possess very powerful absorptive properties, which are especially developed in bone black, in consequence of its fine division. The use of granular bone black (known as "char") in sugar works, and generally for decolourising liquids, is due to the peculiar state of division of the charcoal. For these purposes bone black is used in enormous quantities; it is made in special works. It does not fall within the scope of this work to give a detailed description of the manufacture of bone black, and we must here restrict ourselves to matters of special interest to the colour maker.

The bones are coarsely powdered and freed from fat by boiling; they are then generally carbonised in iron retorts, a number of which are built in a furnace in a vertical position; a valve at the bottom serves to empty the retorts. In a well-arranged apparatus the products of the dry distillation of the bones, chiefly consisting of ammonium carbonate, are collected. The retorts are carefully closed so that air cannot enter, otherwise a portion of the carbon would burn and the black would acquire a grey shade instead of the pure black which it should possess.

Bone black makers who work for sugar factories carbonise by preference the densest bones; these produce a black of the most powerful decolourising properties. This property is of no advantage to the colour maker, who simply requires a very dense black. The bones of young animals, and especially certain bones, contain a larger proportion of cartilage than the hollow bones which produce the best "char". In making bone black to be used as a pigment just those bones should be chosen which are of less value for "char".

On the small scale, bone black can be made by carbonising in crucibles; these hold about 16 kilogrammes of bones, they have a projecting rim, so that when one is placed upon the other it serves as a lid for the first. Piles of these pots are formed, the top one being covered by a well-fitting lid. The piles are heated in a furnace in such a manner that the flames can freely circulate between them. At first a small fire is applied; as soon as dry distillation of the bones begins, which is recognised by pale white flames at the rims of the crucibles, the fire is damped, because the burning products of distillation produce so much heat that the crucibles are soon at a good red heat.

When the flames disappear the fire is maintained for about fifteen minutes longer, and when the crucibles have cooled to some extent they are taken out of the furnace and immediately emptied into a sheet-iron cylinder, in which the charcoal remains until quite cold. If the black comes in contact with air whilst still hot a portion of the carbon burns and the product has little value either as a pigment or for decolourising purposes. When quite cold the black is ground and levigated. The bone ash it contains makes these operations easy. Bone black contains at the most 12 to 13 per cent. of carbon; the remainder consists of bone ash and water absorbed by the hygroscopic carbon from the air.

Bone black which has been partially burnt in the process has a greyish tinge; it may also have an ugly shade of brown: this is the case when it has not been heated to a sufficiently high temperature, so that it still contains some quantity of organic matter. Such black can be made usable by again heating, but it would always be desirable to test a small portion before the whole was ground and levigated; a uniform heating of the finely-powdered black would be attended with difficulty.

Finely ground bone black has numerous uses as a pigment; by a simple process it can be converted into almost pure carbon, which can be used as an excellent black pigment for all purposes for which blacks are employed. Bone black consists of bone ash upon which fine particles of carbon are deposited. Bone ash is easily soluble in hydrochloric acid ; if finely ground bone black is treated with this acid and the residue washed with water until the washings are neutral, a residue of extremely soft and pure carbon is obtained, which has a deep black colour and, in consequence of its fine division, very great covering power.

CHAPTER XL.

MANUFACTURE OF SOOT PIGMENTS.

THE soot which is formed in the incomplete combustion of organic substances containing a large proportion of carbon is a mixture of different substances, of which carbon is the chief; in addition to this, we find in soot almost all the products which result by the dry distillation of the substance from which it is formed. Soot from hard wood contains different compounds to soot from soft wood; again, rosin soot is of a different nature to that obtained in the incomplete combustion of fats. We can thus define soot as very finely divided carbon, mixed with the products of dry distillation.

The soot of hard woods, which contain little or no resin, has a deep black, though dull hue; it is a gritty powder, and has little value as a pigment. Wood containing much resin, such as the wood of the pine, rosin, fish oil, asphaltum, in short, all bodies which are at the same time easily combustible and rich in carbon, produce on the contrary a handsome, glistening soot, which forms a valuable pigment.

In addition to this difference in composition, soot from different sources differs also in the size of its particles. Soot forms light flocks, which adhere to projections in the flues through which the products of combustion pass; the larger the flocks the sooner they are deposited, the smaller they are the longer they remain suspended. The finest particles of soot are called "flying soot," which is very highly prized on account of its fine division.

The manufacture of soot blacks is a very important industry : black printing ink, which is used in such enormous quantities, is made from a soot black; in addition, all the best black paints and lacquers. Soot black was formerly made in the most primitive manner, and is still, to some extent, as will be seen from the account of the manufacture of rosin black. The process used for printing blacks is more rational, but is still capable of great improvements. The principles of a rational manufacture of soot blacks will be briefly indicated.

The Manufacture of Soot Blacks on the Large Scale.—The principle of the arrangements necessary for making soot blacks is very simple. An apparatus is required in which substances very rich in carbon can be burnt at the lowest possible temperature; this apparatus must be connected with suitable arrangements for retaining the soot carried away with the products of combustion. The soot works as at present arranged are developments of the crude arrangements still used in districts in which there is an abundant supply of pine wood containing much resin. The apparatus used in such localities consists of a low masonry flue connected with a long pipe built of wooden boards. In order to give this pipe a rough surface upon which the soot can readily deposit, it is lined in several places with coarse linen, to the projecting fibres of which the soot adheres. In the flue the very resiniferous wood is burnt, especially the roots of pines, which are very rich in resin; these burn with an unrestricted air-supply with a bright flame, but when the air is restricted, they give off a large quantity of a very heavy smoke. The operation is commenced by first making a good fire of dry split wood, the object of which is to heat the flue and thus prevent the deposition of soot in it in the later part of the process. If a deposit of soot formed in the flue it might take fire, the fire would then spread further into the wooden flue and a considerable loss of soot would occur. When the stone-work has become so hot that deposition of soot in it is no longer to be feared, the materials are introduced from which the soot is to be made. As has been said, pine roots are generally used, the chips of pine wood are also used in some districts, and in general such combustible materials as produce a large quantity of smoke.

The combustion in the flue must be conducted with care, it should proceed at the lowest possible temperature, but this should not sink below a certain minimum. If the fire is too strong the greater portion of the carbon is burnt, which would otherwise be obtained as soot; the yield of soot is very small, and generally only "dust soot" would be obtained, without any quantity of "flying soot". On the other hand, if too little air enters, so that the combustion takes place at too low a temperature, a large yield of soot is obtained, but it is of poor quality. Soot obtained by combustion at too low temperatures has not a pure black but a perceptibly brown colour; it has not the flocculent nature of the best product, a small weight of which occupies a very large volume, but is greasy and has a high specific gravity. The lower the temperature used in producing soot black, the further is the chemical change removed from that of ordinary combustion, and the nearer to that of dry distillation. Soot produced with an insufficient supply of air contains a considerable quantity of liquid and solid products of dry distillation, which give it a brown colour and the above-mentioned greasy nature.

In many soot works the air supply is regulated in a most primitive manner : the workman places against the openings by which air is admitted to the burning materials a larger or smaller number of bricks according as the combustion appears to be too rapid or too slow. Unfortunately there is no clear sign to indicate the temperature most favourable for the production of the greatest quantity of soot, the appearance of the flame alone can be taken as a guide. If the flame appears pure white and very luminous and shows at its end no thick black smoke the combustion is very complete, and but a very small yield of soot is to be expected. On the other hand, if the flame continually threatens to go out the air supply is too small and a liberal amount of soot will be formed, but mixed with a large quantity of the products of dry distillation. As far as it can be described in words the flame should appear as follows : The colour should be a murky red, similar to that of the flame of a bad tallow candle ; in shape the flame should be a long drawn-out tongue, from the point of which a thick black smoke is clearly seen. The soot deposits in the long board flues in the form of flocks or dust. The properties of the soot which is deposited in the different parts of the long flue vary according to the distance from the place of combustion, near to which a good black soot is deposited, requiring, however, long grinding with oil or gum solution to form a good paint. At a greater distance the soft and very fine " flying soot " is deposited ; it is the most pure black and is regarded as the best quality. The soot deposited farther on is more and more brown, and has a more greasy nature the greater the distance from the place of combustion.

The flue for the reception of the soot must be made so long that hardly any smoke is perceptible at the end, which it is convenient to connect with a chimney with a good draught, regulated by a damper. In this case the combustion and the the speed at which the products enter the flues are twice regulated.

The material of which the flues are constructed through which the products of combustion are led influences the manner in which the soot is deposited ; the portions nearest to the grate must be of stonework, but when the smoke has cooled to a certain extent different materials can be used in addition to wood ; flues made of sacking stretched over laths are used. The fine flying soot adheres to the roughnesses of the coarse sacking, and is easily loosened by striking the wall of the flue.

As we have said, the arrangements of most soot works are very primitive ; they obtain only a fraction of the quantity of soot which they could produce by a proper regulation of the draught in the combustion chamber and in the flue in which the soot deposits. We shall now briefly describe the conditions which should be regarded in constructing a soot works ; these conditions have not been found to be observed in any works we have seen, in this branch of chemical industry so-called practical experience is alone regarded, and consequently the "practical" soot burner loses large sums, literally up the chimney. A black works can be properly designed by any one who knows sufficient chemistry to understand the process of combustion. Of this most educated men have some understanding, but the practical soot burner appears to have but hazy ideas on the subject, for one occasionally finds the arrangements for making soot in direct opposition to the proper disposition.

For the sake of the uniformity of the product and of safety against fire, the flues in which the soot is deposited should be entirely of masonry ; the joints of the bricks should be smoothly cemented over, so that soot cannot collect in quantity in them. The end of this flue should be connected with a high chimney provided at the top with a well-fitting damper, so that the draught in flues and chimney can be regulated at will or completely stopped. Such an arrangement, perhaps somewhat costly, has many important advantages : it is fireproof, and when once warm remains so for a long time, since bricks are bad conductors of heat. When the flue is hot no water condenses in it; all the water formed in the combustion remains in the form of vapour, and is carried away by the chimney. A further advantage is that it is not often necessary to enter the flue to collect the soot; burning may be continued for a long time, and a large quantity of soot taken at once from the flues or soot chambers.

The soot collects on the walls of the chambers in flocks, which finally become so heavy that they fall off and collect on the floor. The entry to the soot chambers should be through a single iron door, tightly closed whilst the chambers are in use, and cemented round. If this door does not fit air-tight the combustion cannot be exactly regulated by the damper on the chimney. The soot is removed from the chambers by a workman, who sweeps off with a soft brush into a sheet-iron vessel the soot adhering to the walls and lying on the ground. It is extremely important that soot and nothing else should be collected; the brush used for loosening the soot should be so soft that it does not rub off mortar from the brickwork; the workman's shoes should be provided with felt soles, so that no particles are loosened from the floor of the chamber and mixed with the soot. The admixture of the smallest quantity of sand would be extremely harmful in the ensuing grinding of the soot; the mills would be damaged.

The flues for collecting the soot in well arranged works are very similar, but the apparatus used for burning the materials which produce the soot varies greatly according to the nature of the material. Some quantity of the soot black used in the arts is still made by burning pine roots and chips, but for the finer qualities American rosin is largely employed. Earth-wax or ozokerite and the hydrocarbons obtained from petroleum and in the distillation of shale are materials frequently used in making soot blacks of very good quality. For the finest qualities, such as are used for fine printing inks, copper-plate inks and black lacquers, soot obtained by burning fish oils or cheap fatty oils is most commonly used. The great differences in the physical nature of these materials demand the use of different apparatus for their combustion. The space allotted to soot blacks in this work would be far exceeded if the construction of all the forms of apparatus were described; the most important only will be given.

When rosin is the raw material, the combustion can be conducted in flat spoon-shaped vessels placed before a narrow opening into the flue. Fig. 30 shows an arrangement very

Fig. 30.

successful in practice. The spoon-shaped iron vessel, G, stands in a second, G_1, which is filled with water: this prevents the fused rosin from becoming too hot. If the temperature in G rose too high, dry distillation would take place along with the combustion of the rosin, and the soot would be largely contaminated by the products of this distillation. This may proceed so far that in place of the fine flocks of soot a greasy mass is deposited in the flues, consisting of a mixture of soot with products of distillation, from which the soot could be obtained only with great difficulty. The soot and gaseous products of combustion pass through the opening, O, into the flues, R R. This opening is only several centimetres wide, but is nearly as long as the combustion vessel. Above the vessel, G, is a movable iron cover, D, in which are slides by which the air supply is regulated. The cover is only taken off when fresh material is introduced.

The air supply cannot be sufficiently regulated by the slides in the cover; these must be used in conjunction with the chimney damper. The combustion is observed through a thick glass plate let into the cover. At the commencement of the operation slides and damper are completely opened. When the chimney is seen to emit a thick black smoke the flues are filled with the products of combustion, and the speed with which they pass through the flues must be regulated. The current of air is decreased until only a little visible smoke escapes from the chimney, and the flame is no longer white, but a murky red. It should be noted that the first soot obtained from a new works is never of the best quality, the production of this is only gradually attained. The cause of this is that the new masonry is damp and gives up water to the hot gases, so that they are cooled and their velocity disturbed; the soot will also be damp. Thus a greasy soot is obtained. It is of very little use to allow the installation to remain unused for some months after it is completed; the brickwork would only dry superficially, and when it was first heated the presence of the water would become evident.

In order to dry the whole erection to such an extent that the moisture from the brickwork has no effect on the regular course of the operation, it is advisable to commence with materials of little value and to allow a portion of them to be

302   MINERAL AND LAKE PIGMENTS.

lost by the use of a stronger draught than is generally used, so that the flues can be freed from all moisture as soon as possible. This applies equally to all kinds of arrangements for this purpose.

Lamps are used to burn liquid fats and mineral oils. These lamps naturally have a different construction to those used for lighting, which are constructed with the object of burning all the carbon in the oil, so that the temperature may rise as high as possible and the carbon burn at a high

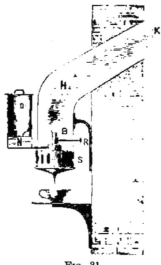

Fig. 31.

white heat. The lamps used in the soot black manufacture are designed to burn only as much carbon as is absolutely necessary to maintain the flame, at the same time the temperature of the flame must be kept low, so that no portion of the soot is again burnt. These lamps have flat burners and are enclosed in a sheet-iron casing provided with an air regulator, which must work very accurately, otherwise air will enter between the joints and the working of the regulator will be deceptive. So that the material to be burnt

MANUFACTURE OF SOOT PIGMENTS. 303

shall not be too strongly heated, which would be accompanied by great loss when mineral oil is used, the reservoir should be placed outside the iron casing which surrounds the burner.

Fig. 31 shows the construction of a soot lamp. The flat burner, B, is placed in the cylindrical sheet-iron mantle, H, which is bent above at not too sharp an angle. The products of combustion are led into a chamber, K, from which they pass into the flues where the soot is deposited. The form of the upper portion of the cylinder is of importance; if this is bent at right angles, soot accumulates on the angle, and when a mass has formed it falls off, and is partly burned in the flame. If the cylinder is given a proper bend no soot is deposited in it, but it is all carried away into the soot chambers. The air regulator, S, is placed at the bottom of the cylinder; it must turn easily. The larger the slits are made by the rotation of this regulator, the more oxygen enters the flame, and the more vigorous is the combustion. A small, well-fitting door is placed in the lower part of the cylinder, through which the wick can be reached; opposite to this a glass plate is inserted, so that the flame may be observed without opening the door. The wick is raised or lowered by means of the screw, R.

The oil reservoir, O, must be outside the cylinder: in the older arrangements it is so placed that the wick sucks up the oil; the workman who is in charge of the lamps must then take particular care that the reservoirs always contain the proper quantity of oil, if he neglects to keep the oil at the proper level the wick is charred; it then sucks up too much oil, which cannot completely burn and is chiefly distilled, so that the soot is oily and cannot easily be treated afterwards. The most diligent and attentive workman who is in charge of a large number of lamps may easily allow one of them to run short of oil. With the oil reservoir above depicted, a new quantity of oil flows only when the level of the liquid sinks below the line, U. As soon as a small quantity is used air enters the reservoir, O, in place of which oil runs out until the opening, U, is again closed by the liquid. This construction of lamp only works well when thin oils are burned, and great care must be taken that the lamps are kept constantly clean.

Lamps of all constructions have some drawbacks, they must always be cleaned, and losses of oil occur in filling. These defects are obviated when, instead of providing each lamp with its own reservoir, a single one is used for a large number of lamps, which are automatically fed from it. In this case the attendant has only to regulate the air supply to the lamps, and to see that the mechanical arrangement by which the oil flows to the different lamps is working properly. When the lamps are automatically fed, the burners must be firmly fixed, and all in the same horizontal plane. A pipe connects each burner with a main pipe running under the lamps, which latter pipe is connected with the reservoir, and this in its turn is connected with another reservoir placed a little higher. In the pipe connecting the two reservoirs is a tap, which is opened by a float in the lower vessel as soon as the level of the liquid in it sinks a little, and which remains open until the level of the liquid has again risen to a certain height. The float in the reservoir connected with the lamps is arranged so that the level of the liquid is slightly higher than the burners. Under the slight pressure oil continually flows to the burners, and it is not difficult to regulate its flow so that all which reaches the burner is burned.

In using a new oil it is not easy at first to regulate the flow so that absolutely all is burned without any dropping off the burner; to prevent the loss of this portion, the lower end of the air regulator is conical, at the apex of the cone is a small tube and under this a vessel in which the unburnt oil is caught. In Fig. 32, S is the air regulator, T the vessel to catch the unburnt oil, L the pipe leading from each burner to the common pipe H, G the vessel in which a float regulates the flow from a larger reservoir, so that the liquid always remains at the same level. When tar oils are used, or thin mineral oils, the pipes which convey the oil to the burners may be made tolerably narrow, but when viscous oils or fish oils are burned, narrow tubes would offer too great resistance to the flow, so that it is always advisable to use fairly wide

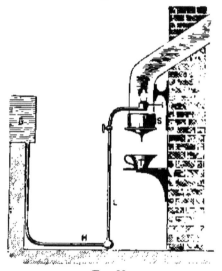

Fig. 32.

pipes. Viscous oils become considerably more fluid at higher temperatures, it is therefore well to place the reservoirs in the same room as the lamps; in winter the temperature of this room is kept fairly high by the burning of the lamps, and thus the oils remain fluid.

In recent years hydrocarbons of very low boiling point have been placed on the market at low prices; these produce a black of very good quality. On account of the low boiling point of these very inflammable liquids, particular care is necessary in burning them: they are very fluid and retain their fluidity at a low temperature. The reservoirs for them should be placed for safety outside the lamp room, and should be closed by an air-tight lid, with only one small opening through which air can enter. When these low boiling liquids are used, particular care must be taken to regulate the flow to the lamp, or large quantities will evaporate without being burnt.

The materials used in the manufacture of soot black differ considerably in chemical constitution. Rosin, animal and vegetable fats, distilled oils and volatile hydrocarbons are used. Each of these substances gives different decomposition products when heated, to which proper regard must be had, since the quality of the black is dependent on them. The more difficultly volatile are the products of distillation, the higher must be the temperature at which the combustion is conducted, otherwise the soot will contain a considerable quantity of these products of distillation. In consideration of the great variety of materials used for making soot black, it is impossible to say exactly in what manner each should be treated; this must be left to the practical experience of the manufacturer.

CHAPTER XLI.

MANUFACTURE OF LAMP BLACK.

THE variety of soot black known as lamp black is the best; it is used for making copper-plate inks and black coach paints. Its price is much higher than that of ordinary soot black; for the best qualities, twenty times the price of an ordinary black is paid. The materials already mentioned are used for making lamp black, fish oils and rancid vegetable oils, and in recent times mineral oil and tar oils. In regard to vegetable oils it is to be observed that it is advisable to use very rancid oil, which gives a larger yield of soot; experience has shown that a very rancid oil requires a larger quantity of oxygen to burn without a smoky flame than a sweet oil. This indicates that a portion of the carbon in the rancid oil is present in such a form that it requires a higher temperature to burn it than is required for the carbon in sweet oils. The use of rancid oils for making soot blacks has thus two advantages, it is much cheaper and produces a larger yield. The only disadvantage, and not a very important one, attending the use of rancid oil is that the free fatty acids in these oils strongly attack the metallic parts of the lamps. This is especially the case with copper and brass; only such portions of the lamps as is absolutely necessary should be made of these metals; all other parts, and in particular the oil reservoirs, should be made of tin plate.

Since the development of the coal-tar industry oils

distilled from coal tar have come into commerce at low prices. These oils consist of carbon and hydrogen and are equally volatile with the essential oils, to which class turpentine belongs. There are two kinds of these tar oils, light and heavy; they differ both in specific gravity and in the boiling temperature, which ranges between tolerably wide limits. When the burning qualities of these oils are examined, they show a great difference in the quantities of oxygen necessary for complete combustion with a white flame without smoke. The more oxygen an oil requires to burn with a non-smoky flame, the more suited it is for the manufacture of blacks. Generally speaking, the difficulty of completely burning these oils is in proportion to their specific gravity and the height of their boiling point.

By the distillation of rosin, oils are obtained which also consist of carbon and hydrogen and form a useful material for the manufacture of soot blacks. The mineral known as earth-wax or ozokerite, a substance intermediate between asphaltum and petroleum, can also be used, but since it is a solid it must be burnt in troughs.

In burning light oils the arrangements can be very simply made; wicks are not required and thus a considerable outlay is spared. In place of the burner, shallow dishes are used into which the oil enters from below, replacing that burnt away. It is necessary to cool these dishes continuously from below, otherwise they would soon become so hot that the greater part of the oil would evaporate without burning, and it would also be very difficult to regulate the flame properly.

Black can be obtained from very resiniferous coal by burning in furnaces with a regulated air supply. The product is, however, generally largely contaminated by ash, and can only be used for common purposes. Very resiniferous lignite gives a better product than coal; the black can be partially separated from the ashes by shaking upon water and then stirring, when the particles of ash sink to the bottom and the light black floats on the top.

However carefully the manufacture of soot black is conducted, it will always contain, in addition to carbon, varying quantities of the products of distillation, partly solid and partly liquid. In consequence of these admixtures the soot will not be pure black, but will show a more or less brown tinge, which is clearly observed when the black is smeared on white paper. At a certain thickness of the layer of soot it will be seen that the colour is not black, but an impure brown. When such soot, as it is taken from the chambers, is chemically examined, it is found that it gives up a large quantity of soluble matter to different chemical reagents. By proper treatment it is possible to remove the admixtures almost completely, so that nearly chemically pure carbon remains. Such pure carbon can be made by boiling lamp black with strong caustic soda solution so long as the liquid is coloured, and when caustic soda can dissolve nothing more, the residue is boiled with *aqua regia* until 'this no longer takes up soluble matter. The black is then washed with water until free from every trace of acid, and the residue dried. By this treatment the soot is converted into a powder of the purest black hue which it is possible to obtain. It is now no longer soot, but chemically pure carbon in the noncrystalline form; heated upon platinum foil it burns to pure carbonic acid without producing smoke or smell. In practice the purification of the soot is not carried to the extent of producing pure carbon; this would be accompanied by a diminished yield of the pigment without increasing the commercial value of the product: the aim of the manufacturer is simply to produce a substance of a pure black appearance from the brown soot. To remove the brown substances present in the crude soot the solvent action of caustic soda solution can be used. The soot is several times boiled in iron pans with strong caustic soda solution in order to dissolve the products of dry distillation. It is superfluous to repeat this operation until fresh caustic soda remains colourless. The treatment may be stopped when the solution acquires a slight brownish colour. When the soot has been purified so far it has lost its brown shade and now appears as a velvety black powder, very soft and very light, and distinguished by extraordinary covering power.

It is not so easy as it appears at first sight to judge of the quality of a black pigment by its appearance. To the unaccustomed eye a pigment may appear to be of an unexceptionable black, which to the expert appears decidedly brown; only long practice can give the eye the requisite keenness. In addition to the test already given—of smearing on white paper—there is another especially to be recommended to the inexperienced for discriminating between pigments. A small quantity of the black under examination is intimately mixed with a white pigment; white lead or zinc white is very suitable. If the mixture has a pure grey shade, the black may be regarded as of good quality, but if it contains brown substances the mixture has an indefinite, dirty shade instead of a pure grey. This is a sure sign that the black requires further purification.

Although caustic soda is now very cheap, its use for the purification of soot black is tolerably costly, because it entails much labour. This process is therefore only used for the finest qualities which are to be employed for copperplate inks and black coach colours. For inferior qualities the method of calcination is used, which produces, when properly carried out, a black of such purity that it can be used for the preparation of even the finest black paints.

Calcination of the Soot.—The substances which give the soot its brown colour are products of dry distillation, and hence are all volatile at a certain temperature; they can be separated from the soot by heating it in the absence of air.

MANUFACTURE OF LAMP BLACK. 311

The temperature necessary completely to volatilise these compounds is fairly high. The soot must be heated to a good red heat to obtain a safe result.

When the soot is heated too quickly or to too high a temperature it undergoes an alteration which affects the quality of the product. By too long or too vigorous ignition the soot changes its flocky consistency to a sandy nature; it will then require much longer grinding with oil to produce a uniform mixture than is the case with the light flocculent soot, which very readily mixes with oil.

The soot is ignited in sheet-iron boxes with a coating to protect the metal from burning. This coating is best made from clay and hair. A very thin paste of clay and water is painted uniformly over the boxes with a brush; when the first coat is dry a second and third are given. When once the metal is completely covered with clay, several coats are given of clay mixed with chopped tow until the layer is several millimetres thick. The coating carefully made in this manner is very durable, and the boxes can be used for a long time, whilst without the coating they would very soon be burnt. Particular care must be given to the manufacture of the boxes themselves; the bottoms must fit very accurately, and should be coated with clay in order to ensure an air-tight joint. The lids also must fit accurately, and when they are closed must equally be made tight by clay.

The soot is at first loosely packed into the box and each portion then pressed with a rammer so that it tightly fills the boxes. In the lid is a very small opening, through which the volatile products can escape. The heating begins quite gently at the back, and proceeds gradually to the front; the boxes are finally brought to a good red heat, at which they are maintained for about half an hour; at this temperature the substances mixed with the carbon are almost entirely volatilised, and the soot acquires its true black appearance. The soot itself attains a red heat in the boxes and in consequence of its loose nature it very readily burns. The precautions given above must be observed in order to protect completely the soot against the action of the air. The boxes should not have the smallest opening besides that in the cover necessary for the escape of the volatile matters; through an opening invisible to the naked eye so much oxygen may enter during the cooling of the boxes that a considerable quantity of carbon will burn to carbonic acid.

To avoid losses through the carbon burning precautions must equally be taken in the cooling of the boxes. When the ignition is finished they are drawn out of the furnace by tongs and placed upright on a stone floor; as the cool air enters, and in contact with the red-hot carbon would burn a portion of it. This can be prevented by a simple artifice: a red-hot coal is placed on the small opening in the lid; this converts the oxygen of the air which enters the boxes into carbonic acid. When all the boxes are taken out of the furnace the doors of the room in which they are placed are opened, so that they are cooled in the draught. Finely-divided carbon takes fire at a temperature far below a red heat, so that the boxes should not be opened until their contents are quite cold. If they were opened whilst hot the soot might take fire.

Pine Black.—Under this name a poor quality of black comes into the market, which is much used for ordinary black paints, etc. The pine black formerly brought into the market was what its name purported; it was made from the roots of the pine in the primitive fashion already described. The black was sold without further purification: it was a soft light powder, very variable in shade on account of the absence of control over the process: it varied from pure black to a dark brown. At present, under the name of pine black many substances are sold, generally without purification, made from the most different materials, frequently from rosin, rosin residues, and other cheap materials.

CHAPTER XLII.

THE MANUFACTURE OF SOOT BLACK WITHOUT CHAMBERS.

A CERTAIN temperature is necessary for the combustion of every substance ; if a cold body is placed in the flame of a candle or an oil lamp it becomes covered with soot, because it cools the flame to such an extent that the carbon is no longer

FIG. 33.

heated to the temperature necessary for its combustion, and is, therefore, separated in a finely-divided condition. Use is made of this phenomenon in a method of making soot black which has many advantages, of which the most important is that expensive buildings are not required, and the manufacture can be conducted in restricted space.

314 MINERAL AND LAKE PIGMENTS.

A convenient arrangement for obtaining black by this process is represented in Fig. 33. The thin-walled hollow cast-iron cylinder is turned smooth on the outer surface, and is surrounded by a sheet-iron cover at the distance of a few centimetres; it rotates in bearings, which, like the spindles, are hollow; cold water is thus led through the cylinder from a tank at a higher level. Below the cylinder are placed, near to one another, smoking lamps, and at the side of the cylinder is a broad brush of soft hair which continuously removes the soot deposited on the surface, which then falls

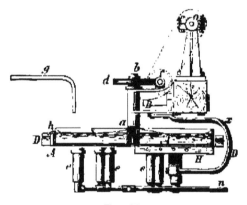

FIG. 34.

over an inclined plate into the collecting vessel. The cylinder is kept in slow rotation by any mechanical arrangement. The lamps are devised not to produce a hot flame, but when this comes in contact with the cold surface of the cylinder, it deposits a ring of soot on the rotating cylinder which is then removed by the brush. The cylinder is kept cooled by the water which runs through it. The soot which is collected shows a tolerably strong brown shade in consequence of the rapid cooling of the flame, which causes the formation of a large quantity of products of distillation; the soot will always require ignition.

It has been proposed further to simplify the manufacture of soot black by this method, by burning coal gas under the cylinder from numerous small openings in a pipe. Coal gas indeed gives a very fine deep black, but the yield is so small that this process could never be adopted with advantage.

The apparatus constructed by Thalwitzer for the manufacture of oil black (Fig. 34) consists of a plate, A, with a rim, a, fastened to a vertical axis, b. This is carried by bearings in B, and is kept in rotation by the cog-wheels, d and f, moved from the shafting, O. The plate is cooled by water supplied by the pipe, g, it flows away by h and the annular vessel, $D.'$ The lamps, e, are connected with the common oil reservoir by n. H is a scraper fastened to B by x.

CHAPTER XLIII.

INDIAN INK.

INDIAN ink consists of purified soot mixed with gum, as binding medium, and a little camphor and musk. It is generally believed that Indian ink is made in China by a process which is still a secret; this opinion is supported by the fact that the ink made in Europe is almost always inferior to the Chinese. It is quite possible that a substance which produces a particularly fine black is used in China to prepare the soot used for the ink, and which we do not yet know or do not use on account of its high price. But it appears that the difficulty of producing an ink equal in quality to the Chinese lies less in the quality of the soot —for by the methods of purification already given we can make almost chemically pure carbon—than in the extremely careful mechanical treatment of the soot with the other constituents of the ink.

The genuine Chinese ink has a peculiar smell, which, when not hidden by musk, distinctly recalls the smell of burning camphor; the camphor tree is a native of China, so that it is not impossible that the soot of the wood of this tree is used for the ink. The soot from camphor alone would be too expensive, so that it is probable that camphor wood is used from which camphor has been prepared, or that camphor soot is mixed with the soot of a cheaper substance. In fact several varieties of Chinese ink show distinctly the smell peculiar to soot produced by fat. The binding material is animal glue, which is most thoroughly ground with the carbon; the secret of making good Indian ink lies in the careful grinding with the binding medium.

According to Piou, who lived for a long time in China and learned the process there used, the black employed for the ink is made from the resins of the pine and other trees. The soot is sieved through silk, boiled with glue and water, and kneaded by hand with a little oil until the mass has become completely homogeneous. It is then left for some days, heated, and pressed into the moulds. This description is not very satisfactory; the products obtained are far inferior to those made by the ordinary European process.

The viscous mass obtained by long treatment of the black with the binding medium is slowly dried, and again ground when it has become thick. When a completely homogeneous paste has been obtained it is made into sticks which are pressed into moulds, and then very slowly dried so that they do not crack. Small cracks always occur; they are filled in by a brush dipped in the thick paste, finally the sticks are entirely or partially covered with gold leaf.

In order to impart to the ink the odour characteristic of the Chinese ink, it is sufficient to use good lamp black, made odourless by extraction with caustic soda, and to add a little solution of camphor in turpentine during the grinding. If it is desired to impart the musk smell which several Chinese varieties possess, a small quantity of a spirit extract of genuine musk may be added.

The size solution is made by long boiling isinglass with water; it must be so concentrated that it forms a strong jelly on cooling. A small quantity of acetic acid is added, so that the grinding may not be hindered by the viscous nature of the size, which the acid prevents from gelatinising. When the mixture has been made by long grinding, the mortar is warmed to 40° or 50° C., whereupon the acetic acid is soon volatilised, and the mass rapidly becomes very thick.

Neutral Tint Black.—A mixture of Indian ink, Chinese blue, and a very small quantity of madder lake forms the colour known as "neutral tint," the shade of which is a peculiar greyish violet. By alterations in the proportions of the constituents different shades of "neutral tint" are produced.

Appendix.—Black Mineral Pigments.

Chrome Copper Black.—When copper chromate is strongly ignited in air and then treated with boiling nitric acid a glittering black compound is obtained, which shows up well when printed on fabrics with albumin. This and all similar pigments are distinguished by great durability.

Chrome Black.—When a mixture of chromium oxide with varying quantities of ferric oxide is strongly heated, pigments are obtained of all shades from dirty yellow and green to the deepest black. This pigment is largely used in porcelain painting to produce a black, which also can be obtained by other but much more costly methods. The best mixture to produce a deep black contains one part of chromium oxide to four parts of ferric oxide.

CHAPTER XLIV.

ENAMEL COLOURS.

An enamel is a glass distinguished from ordinary glazes by a much lower melting point, and generally by opacity. Most enamels are coloured by additions of metallic oxides.

Without going in detail into the art of enamelling, some observations may here be given which appear indispensable for an understanding of the manufacture of enamel colours. The enamels are glasses, but not every glass can be used for an enamel, since the constituents of the glass react with the substances by which the enamel is coloured, and with certain colouring matters quite different shades to the intended might be produced. Of equal importance with the composition of the glass used in the preparation of the enamel is the temperature at which the glass melts. Many colours bear but a low heat; they decompose at a somewhat higher temperature and produce quite the wrong colour. This is especially the case with the enamels coloured purple-red' by gold.

The enamel colours are fixed by mixing them with an easily fusible glass (the flux), applying them to the object to be enamelled, whether metal, glass, or porcelain, and heating the painted article in muffles until the glass melts, and either dissolves the colouring matter or encloses it unaltered. According as the enamel melts as a whole, or the flux alone melts and encloses the colouring matter, so transparent or opaque enamels are produced. The former are really coloured glasses, the latter are glasses in which is enclosed the sintered colouring substance.

Whilst enamels were formerly of importance only for artistic purposes, they have recently attained great industrial employment; not only is earthenware now covered with enamel, and thus made to resist the attack of chemical reagents, but boiler tubes are lined with enamel to prevent the formation of scale.

The colours used for enamels must in all cases be metallic oxides; for yellow, silver and antimony oxides; for red, gold, copper oxide and ferric oxide; for blue, cobalt compounds; for green, copper oxide or chromium oxide. Two operations are required in the production of enamel colours, the preparation of the flux and its fusion with the real colouring substance. In most cases the latter operation takes place simultaneously with the fixation on the enamelled article, but the enamels may be melted and cast into lumps, which are then powdered.

White Enamels.—These are always ordinary crystal glass, to which tin dioxide or potassium antimoniate has been added. Particular care must be taken that only pure raw materials are used, for almost unweighable quantities of iron compounds are sufficient to impart to the enamel a yellow tinge. In almost all cases the glass must be decolourised by the addition of pure pyrolusite (manganese dioxide), which at a red heat gives up a portion of its oxygen to the ferrous oxide contained in the glass, producing ferric oxide, which has much less colouring power. Too large a quantity of pyrolusite imparts a blue tinge to the enamel.

The tin dioxide used in the glaze is made by direct oxidation of tin in air. It has been found that tin burns much more easily in air when it is mixed with lead; 20 to 40 parts of tin are melted with 100 parts of lead, and the alloy heated in shallow vessels in the air; it takes fire at a certain temperature, and it is only necessary to remove continually the layer of oxide in order to oxidise the whole of the metal within a short space of time. The mixture of oxides obtained in this manner is freed from particles of metal by grinding and levigation; when it is fused with the glass, the lead oxide enters into combination, whilst the tin oxide is embedded in the colourless glass. When paste enamels are required the mass is melted in shallow crucibles, poured into water and broken up to a coarse powder, which is again fused. In some cases this operation must be repeated several times to obtain a quite homogeneous product, for tin dioxide is very heavy and sinks to the bottom of the fluid glass; this is readily seen if the fused enamel is allowed to cool in the crucible. When the crucible is broken the lower portions of the fused mass are dense white, whilst the upper are merely milky white. By repeatedly fusing the mass it is endeavoured to obtain it uniform. In fusing the enamel care must be taken that it is heated in a vessel from which the fire gases are completely excluded, since the smallest quantity of coal or of ferruginous ashes coming into contact with the melted mass would injure its colour. If the melting point of the enamel is too high it can be lowered by the addition of a small quantity of pure quartz sand.

A fine white enamel is obtained by using litharge or red lead. The mixture contains 60 parts of quartz sand, 30 of alum, 35 of common salt and 100 of red lead. It is advisable to add a small quantity of finely-powdered talc to the sand. In consequence of the considerable proportion of alumina contained in this enamel it is difficult to melt, and can be heated to very high temperatures without injury to the shade.

When antimony oxide is used a glass free from lead must be taken; lead glass does not give a pure white with antimony oxide. A very good white enamel is obtained by fusing 3 parts of crystal glass with 1 part of sodium antimoniate.

Coloured Enamels owe their colour to metallic oxides. Enamels coloured by metallic oxides, in which the oxygen is firmly united, can be fired without great precaution, but if they contain oxides easily decomposed, great care is required to obtain a fine colour. In enamelling metals a white underglaze is generally used beneath the coloured enamel; it consists of a refractory white enamel. By this means it is more easy to obtain pure colours.

Yellow Enamels are coloured by silver, antimony oxide combined with litharge, or ferric oxide; from the latter a red can also be obtained. To produce a yellow enamel by means of silver the article is first enamelled white at a low temperature; silver oxide is then applied where required, and the article again heated. It then frequently happens that the surface has a metallic appearance owing to the reduction of a part of the silver oxide to metallic silver. When this coating is scraped off the enamel beneath is found to be coloured a fine yellow.

The antimony yellow is obtained by mixing 1 part of antimony oxide, 1 part of alum and 1 to 3 parts of white lead, according to the depth of colour required. These finely-powdered materials are intimately mixed with 1 part of sal ammoniac, and heated in an open vessel with stirring until the yellow colour appears. The vapours of the ammonium chloride indicate the proper temperature for the operation. When this substance is completely volatilised the temperature should not be further raised, or the mixture would fuse.

By means of ferric oxide a fine and durable yellow is produced; the quantity employed must not be too large, or a red colour will result. A very high temperature is required for burning in this colour. The alum used in making the yellow enamel colours serves to prevent the oxides from fusing.

Red Enamel.—Ferric oxide is generally used. When a

purple red is required it is obtained either directly from metallic gold or from purple of Cassius.

Red iron enamel is made by slowly heating 20 to 25 parts of pure ferrous sulphate with 10 parts of aluminium sulphate until all the water of crystallisation is expelled, when the temperature is gradually raised to an intense heat. The shade depends upon the temperature: the higher it is the darker is the colour. Tests are taken from time to time of the mixture, and rapidly cooled. Hot ferric oxide is black, consequently the shade can only be judged with complete certainty in a cooled portion. Although the temperature required for the preparation of this colour is very high, yet it ought not to rise so far that the magnetic oxide is produced, a small quantity of which would cause the enamel to appear dirty red, since it gives to the flux a blackish green colour. The red ferric oxide colour should not be dissolved in the glass of the enamel, but should be embedded in it. Thus, when the enamel is burnt on, the temperature should not rise so high that ferric silicate is formed, otherwise a yellow or even completely black enamel results.

Purple red is obtained by means of gold; formerly purple of Cassius was exclusively employed for this purpose, but an equally bright red can be obtained from gold chloride. The colour obtained from gold will bear only a very low heat; it must be mixed with a very fusible glass, brought on to the article to be enamelled, and heated just sufficiently to melt the mass. The following mixture is used for gold red and other delicate colours: 3 parts of calcined borax, 3 parts of quartz sand and 1 part of chalk. A very small quantity of gold is sufficient to produce a deep red, the amount of gold preparation used for a pale red or a medium purple must be carefully weighed.

Blue Enamels are always coloured by cobalt oxide. Any cobalt compound can be used, since at a red heat the silica of the glass displaces other acids, and produces cobalt silicate. It is most simple to take cobalt nitrate. This salt is readily obtained completely pure, and absolute purity is of great importance, for only thus is a pure blue produced.

Cobalt oxide produces a more or less deep blue when fused with varying quantities of glass. In order to produce shades similar to turquoise and forget-me-not, a white enamel must be used beneath the blue, or bone-ash must be mixed with the blue enamel; this produces a paler colour, which is also opaque.

Green Enamel is coloured by copper oxide or chromium oxide. In the former case 1 part of copper oxide is used to 30 to 50 parts of glass, according to the depth of shade required. The enamel made with copper oxide alone has never a pure green colour; it exhibits a blue tinge. A pure green is obtained by adding a very small amount of ferric oxide, which produces a yellow and compensates for the blue shade, so that a pure green is formed.

Chromium oxide produces a beautiful emerald green without any addition. The enamel may be exposed to very high temperatures without injury.

Green enamels may also be made by mixing blue cobalt enamel with a yellow enamel; an unexceptional shade results, but this method is seldom used, since green enamels are obtained in a simpler manner from copper and chromium oxides.

Violet Enamel.—Manganese dioxide is used exclusively to produce violet enamel; an extremely small quantity is sufficient to colour a considerable quantity of glass. A pure violet shade is only obtained when very pure manganese dioxide is used; the artificially prepared substance should be used to produce the finest violet: the cost is considerably greater than that of the mineral pyrolusite, but equally good results can rarely be obtained from pyrolusite.

Black Enamel.—When a large quantity of ferric oxide, copper oxide or cobalt oxide is fused with a glass, a deep black enamel is obtained. Generally mixtures of these oxides are used; experience has shown that a much deeper black is thus obtained than from any one alone.

Now that enamels are not used exclusively for artistic purposes, but in considerable quantities for earthenware and other technical purposes, the manufacture of enamel colours has attained considerable importance; works already exist occupied almost exclusively with this special branch of manufacture.

By Lacroix' process enamel colours are made which can be applied to porcelain without further admixture with a flux; for example, a very fine blue is made by dissolving 300 parts of pure alumina and 100 parts of cobalt carbonate in nitric acid, evaporating the solution to dryness, igniting the residue and fusing it with 300 parts of quartz sand free from iron, 900 parts of crystallised boric acid and 1,800 parts of red lead. A blue glass is obtained which easily melts, but is difficult to powder; but if poured when fluid in a thin stream into cold water, it forms thin threads, which are very brittle in consequence of the rapid cooling, and may be easily converted into a soft powder.

CHAPTER XLV.

METALLIC PIGMENTS.

METALLIC pigments do not always consist of metals: the name is also applied to compounds which possess a pronounced metallic lustre; mosaic gold, the preparation of which has been previously described, is an example of such a pigment. Another variety of metallic pigment is made by heating finely-powdered alloys: a layer of oxide is thus produced upon the surface of the metallic particles, this layer produces a shade of the colour of the metal or alloy. Metallic pigments are only unalterable when they are composed of metals which are not changed by the action of the air. As a matter of fact, this property is not possessed by any of the metals used for this purpose; even gold and silver are blackened by the action of the sulphuretted hydrogen in the air. This alteration proceeds, however, very slowly when the metal is enveloped by a layer of a binding medium, which is always the case when the metallic powder is used for painting. There are many manuscripts, the initial letters of which have been coloured by gold or silver, in which the metals retain their peculiar lustre after the lapse of centuries.

Metallic paints made from yellow alloys have never much durability: they always contain copper, which readily alters; articles painted with imitation gold soon lose their lustre, and in the course of time become green.

Shell-Gold.—This very expensive artists' paint is made by rubbing gold-beaters' refuse with gum solution upon a stone slab until a completely homogeneous mixture results, which is then rapidly thickened over the fire and generally allowed to dry in small shells. The preparation so made is known as shell-gold : it is sold at very high prices.

The operation of grinding the gold is very lengthy; it considerably increases the cost of the already expensive material. The process may be considerably shortened if the gold is obtained in a very finely divided state by a chemical operation. For this process coins or broken jewellery is heated in hydrochloric acid, and nitric acid gradually added. The gold dissolves in the mixture. If it was alloyed with silver a white precipitate of silver chloride is formed, which is filtered off after largely diluting with distilled water. The solution is boiled for some time to remove excess of nitric acid. A solution of ferrous sulphate is then added; the liquid at once becomes bluish black, and in a short time deposits a brown precipitate composed of chemically pure gold, which is so finely divided that it is almost without the characteristic glitter of gold. The precipitate is filtered off, dried and preserved in stoppered bottles. In order to prepare shell-gold from this gold powder it is simply necessary to grind it with thick gum solution in a porcelain mortar; under the pressure of the pestle the gold rapidly acquires its natural glitter. The grinding need only be continued until the gold and gum solution are uniformly mixed. The mixture so made should be at once filled into the shells, since on standing the gold would soon separate from the gum solution in consequence of its high specific gravity.

Shell-Silver.—Genuine shell-silver can be made by rubbing silver leaf with gum solution upon a slab exactly as genuine shell-gold. In this case also the labour may be considerably lessened by converting the silver into a state of fine division by a chemical process. Silver is dissolved in nitric acid, which must be quite free from hydrochloric acid, or insoluble silver chloride would be formed. In the operation brown suffocating fumes are evolved which attack the respiratory organs; it should therefore be conducted in the open air or under a flue with a good draught. The silver solution, which is generally coloured blue by admixed copper, is diluted with a large quantity of distilled water. A sheet of copper is then dipped into the liquid and rapidly moved about in it. The silver then separates as a dark grey powder; after washing it is chemically pure, and when ground with gum solution it produces genuine shell-silver.

Shell-silver and shell-gold are rarely used in painting on account of their cost; they are chiefly used for illuminated manuscripts. When these metallic pigments are to be used in oil painting, in place of gum they must be ground with a liquid which mixes with boiled oil or essential oils. For this purpose copaiba balsam is to be recommended.

Imitation Silver.—The imitation silver pigments are made from an alloy of tin and bismuth, or from a tin amalgam. The latter is most easily made by melting tin in a porcelain dish and adding one quarter of its weight of mercury; the mixture is then well stirred, and allowed to cool. It quickly solidifies to a crystalline mass, which is tolerably brittle and can be powdered without difficulty. To obtain a substance of a true silvery appearance the amalgam must be converted to a powder of a certain degree of fineness. If the powdering is carried too far the product loses a great part of its metallic lustre, and acquires a dull grey colour.

The bismuth alloy is made by fusing 100 parts of tin, adding 100 parts of bismuth, and then 10 parts of mercury. It is not absolutely necessary to add mercury, but the addition has the advantage that the solid alloy is far more easily powdered. The imitation silver made by this process has a white metallic colour approaching that of genuine silver, but not equal to it, especially in lustre. It is, however, largely used in the industries on account of its low price; for example, for paper hangings.

CHAPTER XLVI.

BRONZE PIGMENTS.

IT would be anticipated from the name that bronze pigments were composed of an alloy of copper and tin; in reality the alloy is composed of copper and zinc, *i.e.*, brass. The bronze pigments are made by a similar process to that described for genuine gold and silver pigments. The waste produced in the manufacture of imitation gold leaf is ground with a solution of dextrine upon a slab until the mixture is uniform and separate metallic particles can be perceived only through a lens. Whilst genuine gold and silver paints are always made in small quantities, on account of the expensive nature of the material, and machinery is not employed, the bronze pigments are in different case; the use of mechanical arrangements is necessary for producing the fine subdivision, otherwise the bronze would be very dear on account of the great cost of grinding. The mechanical arrangements required to divide the alloy are of the ordinary nature, but special machines have been constructed for the manufacture of bronze pigments, by which the metal is far more quickly converted into powder than by means of grinding machinery. These machines consist of metal drums studded on the interior with a large number of fine needles and capable of very rapid rotation. When a metal powder already tolerably fine is brought into these drums, it is rapidly brought to such a condition of fine division as would be attained by hand grinding only by prolonged and laborious exertion.

The raw material for the manufacture of bronze powders is produced in making imitation gold and silver leaf; the waste metal obtained in beating the sheets is used. The employment of this waste has two advantages: the metal is already in very thin sheets and is composed of alloys varying in colour from silver white, through gold, to a bright copper In making leaf metal the waste of each colour is kept carefully separate, so that it simply requires to be broken up to produce bronze powders of different shades.

Before this waste is brought into the drums mentioned above it must be subjected to a preliminary grinding in a mortar with a small quantity of a fatty oil, which serves to bind together the mass. Sufficient oil should be used to give the mass some degree of coherence; if too much oil is added the space between the needles of the drum would be coated with the mass and the process in the drum would require a much longer time. The uniform mixture of oil and bronze waste is then brought upon a wire sieve of the finest possible mesh, and the mass is rubbed through by means of a fine metal brush into a vessel below; thus the larger particles are retained by the sieve and only those which are smaller than the mesh pass through. The product of this process is then brought into the drums, which are rapidly revolved; the small particles of metal are thrown with great force against the side and are converted by the fine points with which it is studded into a very fine powder. The time required for this process depends on the rate of revolution and on the quantity of powder treated at once. The drums are stopped from time to time and the contents examined. When the powder is sufficiently fine it is taken out of the drum. This is most easily accomplished if the drum is arranged to take apart into two halves.

In most works it is usual to free the bronze powder from the admixed oil by subjecting the mass to the greatest pressure that can be produced by a very powerful hydraulic press. The oil which flows from the press is always green, which shows that chemical action has taken place. In consequence of the large surface imparted to the oil it becomes speedily rancid, and then contains free fatty acids which attack copper very energetically./

The use of hydraulic presses may be avoided by removing the oil by means of a solvent. Fatty oils dissolve very readily in carbon bisulphide, but this solvent cannot be used in this case, because commercial carbon bisulphide always contains dissolved sulphur, which would blacken the

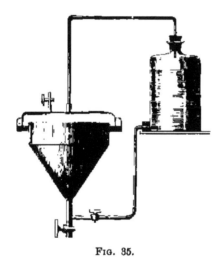

Fig. 35.

bronze powder. Petroleum ether and benzene are very suitable solvents for this purpose. On account of the volatility of these inflammable liquids lights must be absolutely excluded from the room in which they are used, and in order to avoid loss of solvent the bronze powder must be treated in closed vessels. The safest plan is to use a special apparatus of simple construction to dissolve the oil. Fig. 35 shows the construction of an arrangement suitable for this purpose. It consists of a cylindrical vessel of tin plate surrounded by a rim into which fits the edge of the cover. When the lid is placed on and the rim filled with water, the contents of the vessel are closed in air-tight and cannot evaporate. The lower portion of the vessel is conical, and is joined to a pipe in which is a tap, and which communicates at the side by a tube with the glass vessel in which the solvent is contained; another tube connects the neck of this vessel with the cover. When this apparatus is used for extracting the oil from bronze powder, a filter of strong blotting-paper is placed in the conical portion of the vessel, care being taken that it fits accurately so that it is not torn by the weight of the bronze. The oily bronze powder is placed on this filter, the cover set on, and the rim filled with water. By opening the tap attached to the solvent reservoir the liquid is allowed to enter the cylinder from below, the air in the latter passing through the tube in the cover to the reservoir. After some hours the oil is dissolved, the tap in the cover is then opened and the liquid run off by opening the lowest tap. If the bronze powder is not quite free from oil after one treatment with the solvent, the operation is repeated with a fresh quantity.

When free from oil the powder and filter are removed from the apparatus and dried; the dry mass forms a solid cake, which is broken up in a mortar and by a little grinding changed into a fine powder.

The colour of the bronze powder is the same as that of the alloy used, but the shade is always rather paler than that of the coherent metal; regard must be paid to this circumstance in making a bronze of a determined colour: the alloy employed must have a rather deeper colour than the shade the bronze is to possess.

The manufacture of leaf metal and bronze powder is frequently conducted in the same works, which also often prepare the requisite alloys; we, therefore, give a few examples of the composition of the alloys which produce certain shades.

The more zinc the alloys contain the lower is their melting point, the greater their brittleness and hardness and the paler their colour. An increase in the copper causes the colour of the alloy to approach more nearly to that of gold, and increases the malleability, a property useful in making leaf metal, but not desirable in making metallic powder. A zinc copper alloy which contains between 1 and 7 per cent. of zinc has an almost pure red, or even a dark red colour; an alloy containing 7·4 to 13·8 per cent. of zinc has a pure golden yellow colour, between 16·6 and 25 per cent. of zinc a yellow appears. An increase of the percentage of zinc above this point produces the colour of brass; it is noteworthy that an alloy containing still more zinc, 33 to 41 per cent., again shows a reddish colour, which is most developed when the alloy contains equal parts of zinc and copper. If the zinc is increased still further the shade gradually goes over to white; this change is already observed in an alloy containing 51 per cent. of zinc, which shows a pure golden yellow colour, and is very brittle. When the zinc rises to 53 per cent. the colour is reddish white; at 56 per cent. it is yellowish white, at 64 per cent. bluish white, and between 75 and 90 per cent. the alloy is bluish grey.

The alloys for bronze powders of different shades have the following composition, according to R. Wagner:—

	Copper, per cent.	Zinc, per cent.
Pale yellow	83	17
Red	94–90	6–10
Deep red	100	—

Bronzes from English, French and Bavarian works contain the following percentages of copper:—

ENGLISH BRONZES.

Orange	9·82 per cent.
Deep yellow	82·87 ,,
Pale yellow	80·42 ,,

French Bronzes.

Copper red	97·32 per cent.
Orange	94·44 ,,
Pale yellow	81·29 ,,

Bavarian Bronzes.

Copper red	98·92 per cent.
Violet	98·82 ,,
Orange	95·80 ,,
Straw yellow	81·55 ,,
Speiss yellow	82·34 ,,

In each case the remainder of the alloy consists entirely of zinc.

Alloys containing from 1 to 35 per cent. of zinc are only malleable in the cold. The malleability is at the greatest with a content of zinc between 15 and 20 per cent.; such alloys are the most suitable for making leaf metal. Alloys containing between 36 and 40 per cent. of zinc may be hammered either cold or hot, whilst the former alloys become brittle on heating. When the percentage of zinc is still further increased the malleability decreases. The most brittle alloys contain 60 to 67 per cent. of zinc.

The alloys are made in a furnace with a good draught, for copper liquefies at a very high temperature. To prevent loss of copper by oxidation the molten metal should not come in contact with air; it should be covered by a layer of red-hot coal, which prevents oxygen from reaching it. When the copper is completely melted, which is ascertained by stirring with a piece of wood, the whole of the zinc is added. Some skill is required in this operation, otherwise a large proportion of the zinc will be volatilised, and the vapours will burn when they come in contact with air, in which case dazzling bluish white flames are seen over the crucible. The best method is to throw the zinc into the crucible and immediately stir it into the molten metal with a wooden rod. The products of the dry distillation of the wood, which are given off in great quantity at this high temperature, keep the air from the surface of the metal and prevent the oxidation of the zinc vapours. The zinc is thoroughly mixed with the copper by stirring with the wooden rod, the crucible is then slowly cooled, with the precaution that the surface of the metal is kept covered by red-hot coals so long as the metal is fluid. When sufficiently cool the metal is poured into shallow iron moulds, in which it quickly solidifies; it is then rolled into sheets, which may be converted into thin leaves by hammering in a similar manner to that in which the gold-beater makes gold leaf.

To obtain bronze powders of different shades alloys of different colours may be used; the bronze powders may also be shaded by two methods—either by adding certain colouring matters of very great colouring power or by partially oxidising the finely divided metallic powder. In the first process the finely ground colouring matter is mechanically mixed with the bronze powder. The use of manual labour would involve a great loss of time; even when quite small quantities of bronze and colouring matter are mixed in a mortar it is necessary to grind diligently for a very long time before a mixture of homogeneous appearance is obtained. In working on a somewhat larger scale it is therefore advisable to adopt mechanical mixing arrangements. A very simple apparatus suffices. A sheet-iron cylinder is used which can be revolved, and provided with a well-fitting slide. In this cylinder are placed the bronze powder and the colouring matter until it is about half full; then, after tightly closing the slide, it is set in slow rotation, which is continued until a test taken out shows a uniform colour.

When bronze powder is slowly heated in a shallow vessel the colour begins to darken at a temperature not much above the boiling point of water. In consequence of the fineness of the particles of the metallic powder the copper readily takes up oxygen, and is superficially converted into copper oxide. This oxide is of a darker colour, and thus by this method the shade of the bronze can be deepened as desired. This simple operation requires a certain amount of practice to produce a product of a determined shade. The desired result is most safely attained when the bronze is spread out quite uniformly in a thin layer upon a metal plate, which is gently heated from below. The powder soon begins to darken; by cooling the plate the progress of the oxidation may be arrested at any moment.

Recently bronze powders have come into the market showing all possible colours in the deepest shades, by the aid of which very remarkable colour effects can be produced. These bronzes are made by dissolving an aniline dye in a little alcohol, pouring this solution over the powder, and mixing the dye uniformly through the whole of the bronze by working the mass for a sufficient length of time. In this way bronze powders are produced which possess a green, red, blue or violet lustre, according to the colour of the dye used. These colours with metallic lustre can also be produced by bronzing the article with a white (zinc) bronze, and then coating it with a varnish in which the required aniline dye is dissolved. A bronze with a fine golden red glitter is produced by applying a golden yellow bronze and then a varnish in which a little aniline red is dissolved. It should be observed here that these effects, produced by a coat of varnish in which an aniline dye is dissolved, only turn out well when the dye is used in very small quantity, for these colours are the strongest with which we are acquainted, and in colouring power far surpass cochineal carmine, which is renowned for this property.

When bronze is coloured by dyes the most varied shades can be obtained with a metallic lustre. According to Conradty a very fine blue bronze is obtained by boiling white bronze for some hours with a weak alum solution, washing and drying, and then mixing in a mortar with a strong solution of aniline blue in alcohol until the solvent has evaporated. This operation is repeated until the desired depth of shade is obtained. The bronze is then washed with pure water. Conradty also recommends that the coloured bronze should be ground with a little petroleum, and then exposed to the air to allow the petroleum to evaporate. This operation, for which no chemical reason can be given, is quite unnecessary. If other dyes or mixtures of them are used in place of aniline blue, bronzes of corresponding colour are obtained.

However handsome are the bronzes coloured by this process, nearly all have the disadvantage that the colours have little permanence, and quickly fade when exposed to light. This is especially the case when the bronzed article is coated with an oil varnish; if, however, a spirit varnish is used, or indeed any varnish composed of a resin and a volatile solvent, the colour of the bronze, protected by the layer of resin, remains quite unaltered for a long time.

Electrolytic Copper Bronze.—Electrolytically precipitated copper may be used as a bronze pigment; it is most simply made by adding pieces of metallic zinc to a solution of copper sulphate free from iron and violently shaking the flask for a long time. The liquid becomes warm, and the copper separates in the form of a very fine precipitate, which is collected on a filter and washed with air-free water (boiling water is best) and then quickly dried. The upper portions of the precipitate in the filter, which are exposed to the air, have generally a brownish colour due to the incipient oxidation of the finely divided metal. They are removed, and the lower portions show the characteristic colour of pure copper.

In the same way silver can be precipitated from a solution of silver nitrate in a finely divided state, but the particles of the silver powder are so very small that they reflect very little light, and consequently the powder has an unsightly grey colour. When a surface painted with this silver is rubbed gently with a hard body, the metallic lustre appears.

Tungsten Bronze Pigments are expensive and rarely employed. They are obtained by fusing sodium or potassium tungstate in a porcelain crucible and gradually adding tungstic acid until the mass has an acid reaction. Tin dioxide is then added in quantity sufficient to neutralise the tungstic acid; the mass is cooled and finely powdered. According as potassium or sodium tungstate is used, a violet or reddish pigment is obtained which exhibits the peculiar metallic lustre of a bronze powder.

Still more costly is vanadium bronze, which is made by adding ammonium vanadate to a solution of 2 parts of copper sulphate and 1 part of ammonium chloride with continual stirring, until the precipitate no longer re-dissolves on stirring. The liquid is then heated for several hours to about 35° C., when vanadium bronze separates in golden yellow scales. These are collected on a filter, washed and dried. When ground with oil or gum solution they can be used as a red gold bronze. The colour is unaltered by the air.

CHAPTER XLVII.

VEGETABLE BRONZE PIGMENTS.

UNDER this apparently contradictory term, substances come into the market which produce a peculiar metallic lustre. When applied under certain conditions, the appearance is similar to that produced by real bronze. The vegetable bronze pigments are lakes as pure and free from foreign admixtures as possible. The lakes obtained from red wood or logwood can be used for this purpose.

From the red woods (*see* p. 384) a magnificent bronze pigment can be obtained, which is either pure golden yellow or possesses a greenish golden metallic lustre not unlike the colour of the wing cases of the rose-bug. To obtain either shade a pure lake is first made by extracting red wood with boiling water, adding a little carbolic acid to the decoction (0·01 per cent. of the quantity of liquid) and allowing to stand for several weeks. The liquid is syphoned off from the deposit, heated, and alum added equal in quantity to 10 per cent. of the wood used. The mixture is then allowed to stand for about a week, the precipitate is filtered off, washed, and, if necessary, dried. If the bronze is to be used in the form of water-colour, the precipitate is dried to a thick paste and mixed with about 10 per cent. of its volume of thick gum solution, so that a viscid mass is obtained which can just be applied with the brush. When the coating is made so thick as to hide the surface of the bronzed article, it has when dry the golden green colour.

In using this lake to prepare a pigment similar to gold bronze it must be almost completely dried, and then mixed with the liquid obtained in the following manner: White soap is melted on the water-bath with the smallest possible quantity of water, and when completely dissolved the same quantity of white wax is stirred in, finally water is added, so that the cooled liquid has the consistency of a moderately thick varnish. When this liquid is ground with the requisite quantity of the still damp lake and the mixture applied to paper, wood, or leather, and after drying rubbed with a glass ball, it gradually acquires a very fine golden bronze colour. This method of bronzing is largely used in the manufacture of wall papers and for colouring fancy leather. To protect this coating against the action of water, it should be varnished when dry.

These bronze pigments may also be used in varnish. The lake is then completely dried and ground with varnish in such quantity as to give a thick mass which can just be brushed on.

The vegetable bronze is obtained from logwood in a similar manner, a solution of stannic chloride being generally used to precipitate the lake when a deep bronze is required, and alum for a pale gold shade. By using mixtures of the two salts intermediate shades are produced. The precipitate produced by alum may be shaded by means of potassium bichromate. Hæmatoxylin forms with chromium oxide a deep bluish black compound which has such intense colouring power that it is used to colour writing ink. The ink is prepared by adding a little potassium bichromate to a decoction of logwood. If a very small quantity of this dark precipitate is mixed with the lake precipitated from logwood extract by alum solution, colours are produced possessing the peculiar metallic lustre and a shade depending upon the quantity of potassium bichromate added. The addition of this salt must be very carefully made, as a very small excess is sufficient to render the colour so dark that it is useless as a bronze. The shade of the precipitate depends upon the concentration of the liquid and other conditions, so that it is impossible to give precise quantities. In practice, the safest and most convenient method is to dissolve the bichromate in a large quantity of water and add very small quantities of the dilute solution to the logwood extract along with the alum. After each addition a portion of the precipitate is rapidly mixed with the above-mentioned solution of soap and wax, to which a little size has been added, and then spread upon paper. If the desired shade has not yet appeared, a little more bichromate is added and another test made, and the process repeated until the proper shade is attained.

Pigments for colouring wall papers and fancy leather are not easily made which produce such fine effects at so small a cost as the vegetable bronze pigments, which deserve the greatest regard from the colour maker and leather and paper manufacturers.

Appendix—The Brocade Pigments ("Brocatfarben").—Under this designation powders have been recently introduced characterised by a strong metallic or glassy lustre, and very suitable for certain purposes, such as the manufacture of wall papers, since they enable remarkably fine effects to be produced. They consist of mica in a tolerably fine state of division. Mica is a mineral which occurs frequently in nature and which very readily splits into thin sheets; it occurs in various colours. In thin sheets mica is colourless and completely transparent; in somewhat thicker pieces it generally shows a distinct metallic lustre similar either to that of gold or silver. When ground to fairly fine powder, it has the same gold or silver lustre, and gold or silver brocade colours are distinguished in commerce. They are made by grinding mica which has been previously sorted according to its appearance. The powdered mica can be sold as powder or may be mixed with a binding substance in order to be ready for use. Gum Arabic is generally used as the binding medium, but it can be replaced by the cheaper dextrine. When printed upon paper the brocade pigments produce the effect of a bronze, and in addition to their cheapness they have the great advantage of being completely unaltered by the air.

PIGMENTS OF ORGANIC ORIGIN.

CHAPTER XLVIII.

LAKES.

The pigments commercially designated lakes generally consist of an organic colouring matter united with a metallic oxide. In isolated cases other compounds of colouring matters are included under the term, such as indigo sulphonic acid, and yet more rarely pure colouring matters are included, for example carthamine red. But the overwhelming majority of lake pigments are compounds of a colouring matter with the oxide of a metal. For this purpose the oxides of tin, lead and aluminium are commonly used.

Lakes are generally made by mixing the solution of the colouring matter with the solution of a salt of the metallic oxide, and precipitating the oxide by an alkali. The colouring matter is separated along with the oxide and forms with it a substance known as a lake. It has not yet been decided whether the lakes are true chemical compounds of the colouring matter with the metallic oxide, or whether the colouring matter is simply held fast by the surface attraction of the finely divided oxide. In favour of the latter view is the fact that a larger or smaller quantity of the colouring matter can be united with a given quantity of the oxide.

The lakes vary greatly in durability; some, such as the

madder lakes, can be counted among the most durable pigments, whilst others have very little permanence, as for example, the logwood lakes.

The majority of the colouring matters used in the preparation of lakes are of vegetable origin, but several are derived from animal sources. The properties of the colouring matters used for this purpose vary greatly; each material demands a special treatment for the production of lakes. Therefore, in describing the methods by which lakes are made, we shall give first the properties of the colouring matter in question, and then proceed to the preparation of the lake. In the preceding portion of the book the pigments have been arranged according to their colour; the lakes will be described in the same order. The yellow, red or green lakes of different origin will be treated in the same section in order to facilitate reference to any particular pigment.

White lakes are not known, neither are there lakes which can be described as black. With these exceptions lakes of all colours can be made and also of all shades. As an example of this madder lake may be given, which is found in commerce in a great variety of shades; it can be made from the palest rose red to the deepest purple red, or, more properly, madder red, which is a characteristic shade. The shades of any lake are obtained by mixing in a white pigment just as in the manufacture of mineral pigments. Whilst, however, in the latter case the white pigment must often be mechanically mixed with the colour, the shading of the lakes is accomplished in their preparation. Generally speaking, a pale shade of a lake is obtained by increasing the amount of the salt, the oxide of which is used to precipitate the colouring matter. The oxides in question are white (with the exception of lead oxide); thus in the pale shades a small amount of colouring matter is precipitated upon a large amount of oxide, whilst in the deep lakes the reverse is the case, a large quantity of colouring matter is precipitated upon a small quantity of oxide, and the shade thus appears very deep.

Apart from the nature of the colouring matter contained in a lake, and considering only the metallic oxide with which the colouring matter is united, it appears that lakes containing lead oxide have little durability. The combination between the colouring matter and the metallic oxide in a lake is so loose that it is easily destroyed by sulphuretted hydrogen; if a lake containing lead is exposed to the action of air containing that gas the colour will, in the course of time, inevitably blacken. The lakes containing tin are also susceptible to sulphuretted hydrogen; in air containing but a trace of this gas they quickly lose their brilliance and in time are quite discoloured.

Alumina is not affected by sulphuretted hydrogen, and must thus be regarded as the most suitable oxide for the preparation of lakes. It is generally applied in the form of alum, in the selection of which great care is necessary if fine colours are to be produced. Commercial alum frequently contains ferric oxide; when the alumina is precipitated simultaneously with the colouring matter the ferric oxide is also thrown down, and is mixed with the lake, its dark colour influencing the shade of the lake greatly to its disadvantage. The effect of the ferric oxide upon the colour of the lake is so important that it is impossible, for example, to obtain a pale red lake with alum containing iron. To avoid the bad results given by alum containing iron it should be examined before it is used, and if it is found to contain any considerable quantity of iron it should be rejected for this purpose.

After alum a solution of stannic chloride is most commonly used to precipitate lakes, but it must be free from stannous chloride; it produces as a rule darker lakes than alum. Care must be taken that the stannic chloride solution is free from iron.

The ordinary process for making lakes is very simple: a clear aqueous solution of the colouring matter is obtained; to this alum or stannic chloride solution is added in proportion to the amount of colouring matter dissolved by the water: the metallic oxide is then precipitated by an alkali. Sodium or potassium carbonate or caustic alkalis may be used; ammonia is very suitable for this purpose, since it is free from iron. The precipitant must be cautiously added, it is introduced in drops when the greater part of the colouring matter has been precipitated; an excess of the alkaline solution would have considerable effect on the shade of the precipitate. The precipitated lake rapidly settles to the bottom of the almost colourless liquid; the settling is particularly rapid when stannic chloride has been used. The liquid is then drawn off, the pasty residue brought upon a strainer, washed several times with water, and dried in the air or in stoves.

When dry, a properly prepared lake forms a mass of little coherence, and can be readily ground to a soft powder, which may then be ground with oil or gum solution to produce oil or water paints. At the same time the materials are added to the pure lake which are used to shade its colour; white pigments are generally employed for this purpose—the paleness of the shade is proportional to the amount of white pigment introduced. Certain lakes are shaded by small quantities of other colouring matters; for example, the admixture of a small quantity of a blue pigment with a pure red lake produces a much deeper colour inclining to purple. If a small quantity of a red or blue colouring matter is added to a yellow pigment there is produced a deeper yellow inclining to orange, or a yellow inclining to green, in proportion to the quantity of the added colouring matter.

It should be observed that it is of great importance for the colour-maker to have a practised eye, sensitive to fine differences of colour. He may then produce colours faultless in shade, for if by chance a colour does not turn out well it may be improved by judicious admixtures. It is quite impossible to give definite rules for the mixing of colours; the estimation of a shade cannot be taught in words, it demands long practice. By the use of a system, provided the user has normal eyes, very slight differences of colour can be readily estimated. A scale of colours should be made from pigments of pure and definite shades, in which scale successful mixtures are gradually inserted, so that a series is produced in which the separate pure colours blend regularly into one another. Then, when it is required to produce a pigment corresponding to a certain shade on the scale, with a little practice the shade can be readily produced, for the scale indicates which colour predominates.

The so-called sap-colours are generally characterised by very pure shades. They are produced by decomposing a lake by an acid or a strong base, evaporating the solution of the colouring matter so obtained at a gentle heat and mixing with gum, starch, or some other thickening material to produce a mass of such consistency that it can be formed into balls or sticks.

The decoctions of the dye-woods always contain other substances in addition to the colouring matters, which are precipitated with the colouring matter, and somewhat injure its shade. When such a lake is decomposed the colouring matter is obtained in a purified form, thus showing its full beauty. This solution of the purified colouring matter might be again precipitated with a salt solution and lakes be produced which would be much brighter than the original, but the losses incidental to this double precipitation would be so great that the lake would be made much too expensive.

CHAPTER XLIX.

YELLOW LAKES.

YELLOW colouring matters are widespread in nature; from them are obtained many yellow pigments used in dyeing and in the manufacture of colours. The colours produced by the yellow vegetable colouring matters are not particularly bright; thus the yellow lakes are used chiefly to produce cheap pigments, whilst the inorganic pigments are used for bright yellows.

Dutch Pink.—Several species of buckthorn (*Rhamnus*) contain a yellow colouring matter—xanthorhamnin—which is obtained pure by extracting the yellow berries with hot alcohol. On cooling, the impure colouring matter separates: by repeated recrystallisation from alcohol it is obtained in the form of crystalline needles, which are soluble in water and alcohol.

The yellow lake known as Dutch pink is prepared from yellow (Persian) berries, by boiling the crushed berries with water and mixing the extract with a solution of alum. The lake is then precipitated by the addition of powdered chalk. As a rule, 500 parts of water are used to 100 parts of berries, 20 parts of alum are added to the decoction, and the mixture poured upon 75 parts of finely powdered chalk. The liquid is decanted off, the residue filtered, washed and dried. Commercial Dutch pink is made from a mixture of the decoctions of yellow berries, quercitron bark and turmeric, to which the alum solution is added, and then chalk. The precipitate is made into conical lumps, which are sold as Dutch pink, and used for ordinary painting and for colouring leather.

Weld Lake.—The dyers' weld (*Reseda luteola*) contains a yellow colouring matter formerly much used in dyeing; a yellow lake can also be obtained from it. When weld is boiled with water a deep yellowish green decoction is obtained, from which yellowish green flocks separate on cooling. The lake can be obtained from all parts of the plant except the root, the flowering shoots giving the largest yield of colour; the lake is generally precipitated by alumina. Equal parts of weld and alum are boiled with water until the latter is dissolved and the liquid is coloured deep yellow; the hot solution is quickly filtered through a thick linen cloth and soda solution gradually added in small quantities with continual stirring so long as effervescence follows.

Alum entirely free from iron is required to produce a bright yellow lake; a small quantity of iron has great influence on the shade. Weld contains more or less tannin, which gives very dark blue or green compounds with iron salts, the production of a very small quantity of which suffices to convert the yellow shade into an ugly dirty colour.

Chalk may be used instead of soda to precipitate the alumina; in this case the precipitate will contain calcium sulphate in addition to the alumina compound of luteolin.

Weld lake may be used in oil, or size, or as a water-colour.

Gamboge Lake.—Crude gamboge is the dried sap of certain East Indian trees; it is used as an artists' colour, but only in water; for use in oil it is converted into a lake. A particular treatment is necessary to obtain a handsome lake. Gamboge is treated for several days with water and the soft mass ground on the slab or in a mill until it forms a uniform paste, which is mixed with water to a thick liquid, which is then put through a fine sieve. A hot solution of alum is then added in the proportion of three parts of alum to one part of gamboge, the mixture is boiled in a wooden vessel and two parts of nitric acid added whilst well stirring; finally a dilute solution of potash is added in small quantities until the liquid solidifies to a jelly, which is spread out in thin layers on filter cloth, well washed with water and dried at a gentle heat.

Prepared Gamboge.—Crude gamboge cannot be used in oil painting because it never gives a uniform shade, but when the colouring matter is freed from admixtures and ground with oil, it gives a deep durable colour of great beauty.

The colouring constituent of gamboge is a resin which is fairly soluble in strong alcohol, whilst the impurities are insoluble. The pure colouring matter is obtained without great expense in the following manner: Coarsely powdered gamboge is placed in a large flask with strong alcohol; the flask is well closed, placed in a warm position and repeatedly shaken; the deep yellow solution is carefully poured off from the sediment, water is then added, when the colouring matter separates in flocks. Only sufficient water should be added to precipitate all the colouring matter; test portions of the liquid should be examined from time to time, the alcohol can then be recovered by distillation and used again. The colouring matter separated from the alcoholic solution forms a hard mass on drying; it must be very finely powdered before it is ground in oil.

There are many other yellow lakes which can be obtained by adding alum to an aqueous decoction of the colour-bearing material and neutralising the liquid. Such decoctions are obtained from fustic, quercitron (*Quercus tinctoria*), young fustic (*Rhus cotinus*), the root of barberry (*Berberis vulgaris*), annato (the fruit of *Bixa orrellana*), turmeric (the root of *Curcuma longa*), etc. The yellow lakes obtained from these sources are seldom used in painting, the mineral pigments are preferred on account of their finer appearance. These colouring matters are largely used in dyeing to produce shades varying from yellow to brown, the majority are also used in calico printing.

Fustic Lake.—Of the colouring materials mentioned above fustic produces a lake which is so handsome that it deserves larger use in painting than it has yet found. "Old fustic" is chipped immediately before use and extracted with boiling water; whilst still hot the extract is mixed with a hot solution of alum; on cooling a fine yellow precipitate is formed, which when dry is frequently sold under the name of Dutch pink. (Genuine Dutch pink is made from a decoction of yellow berries; the majority of the pigments sold under this name are made from decoctions of mixtures of the different yellow dye wares.)

To obtain the best lake from fustic the alum used must be completely free from iron, otherwise only a dirty green lake is produced. Since the least trace of iron damages the shade of the colour, fustic lake is more frequently made by means of lead oxide. To obtain the lead fustic lake the decoction is allowed to stand for several days, so that the morin, which together with maclurin forms the colouring principle of fustic, may separate; the liquid is filtered from the precipitate, and lead acetate solution added which has been boiled with litharge to saturation, so that the most basic acetate is produced. The lead fustic lake has a deep yellow colour; it is well adapted for use in painting, but, like all lead pigments, is not specially permanent in air. By admixtures of levigated chalk or starch paler shades of the dull yellow lake are obtained.

Quercitron Lake.—Lakes are seldom made from quercitron alone, although they possess a deep yellow colour; they are more frequently employed in dyeing when they are produced directly upon the fibre. Quercitron lake can be made by treating the aqueous decoction of the ground bark with tin crystals and a little alum solution. With tin crystals alone a darker yellow lake is obtained than when alum is used at the same time. Different shades may thus be produced.

Extract of quercitron bark comes into commerce as a greyish yellow powder under the name of flavine, which has great colouring power, and may be used with advantage in the place of aqueous extracts of the bark.

Since the discovery of the aniline dyes the applications of all lakes have been largely extended, since the dyes enable brighter shades to be obtained without the many processes necessary when natural colouring matters are used, and do not require so much skill on the part of the workman. This applies not only to yellow colouring matters, but to the many other colouring matters formerly used by the dyer. The principal reasons which have so quickly brought the new dyes to the front lie in their greater beauty and in the diminished labour required in their use.

Purree or Indian Yellow must not be confounded with the yellow cobalt pigment known also under the latter name. It is a compound of magnesia with an organic acid—euxanthic acid. It is obtained from the urine of cows fed upon mango leaves. It comes into commerce in lumps weighing about 50 to 60 grammes, which are dark brown on the exterior, but on fracture show a fine orange yellow colour. Indian yellow is little used, and is not likely to attain importance. Many cheaper colouring matters equal in shade are known.

The Colouring Matter of Saffron.—The dried stigma of the flowers of the saffron, a species of *iris*, has been long used for colouring foods. It contains a very handsome yellow colouring matter of a characteristic shade; formerly this was largely used in silk-dyeing, and especially for dyeing glove leather. It is practically now no longer used; considerably cheaper aniline dyes produce an equally good shade. On account of its high price (1 kilogramme of saffron contains about 60,000 stigmata) this material cannot be used for lakes, although they leave nothing to be desired in fineness of shade, but they have no great permanence.

The Colouring Matter of Gardinia Grandiflora.—The fruits of this plant, which are imported from Southern Asia, contain a beautiful dark yellow colouring matter, which, according to Rochleder, is identical with the colouring matter of saffron. It is readily soluble in water; lakes of various shades can readily be obtained by the addition of salts to the aqueous solution. Alum gives a pure yellow, lead acetate a yellowish red, and stannous chloride a dark orange red lake.

CHAPTER L.

RED LAKES.

COCHINEAL AND CARMINE.

THE purple of the ancients was obtained from the shellfish *purpura*; it was so costly that it was regarded as an attribute of royalty. According to history the Tyrians discovered this colour; at any rate they were able to use it for dyeing. In the writings of the ancients Tyrian purple garments were regarded as the greatest luxury. The art of obtaining a purple from this source was lost; more recently it has been re-discovered, but the colour is found in no way to correspond with what we regard as a bright shade.

When the Spaniards conquered Mexico they found an insect cultivated which produced a splendid red colouring matter. This insect, the cochineal, is parasitic upon certain cacti, especially *Cactus coccinellifer* and *Cactus opuntia*. The females only are used as colouring material; the males are very small, and much fewer in number. The cultivation of the proper species of cactus and of the cochineal has spread to most tropical countries. The females, which are attached in enormous numbers, are brushed off and killed by heating upon hot plates. They then appear as grains of the size of millet seeds, with a wrinkled surface covered with a silvery grey dust. This variety of cochineal is the best. Black cochineal also comes into commerce; it has a brownish black colour. It is produced by killing the insects in boiling water, by which the grey dust which covers the living insect is removed. Cochineal produces when ground an ugly reddish brown powder.

The animal nature of cochineal cannot be recognised by the naked eye, it is seen on examination under the microscope; this has given opportunity for the most incredible adulterations of the costly substance. Cases have been known in which a paste chiefly composed of flour has been pressed into the form of cochineal, the grains coloured by some cheap red colouring matter, powdered with the dust from boxes in which cochineal has been packed, and placed upon the market as cochineal. The expert would not be deceived by this gross fraud, but another method of adulteration is far more difficult to recognise; genuine cochineal is brought into commerce from which the greater part of the colouring matter has been extracted, and which has been again powdered with the grey cochineal dust. Such extracted cochineal does not produce the same brownish red powder as the fresh substance.

In the cochineal insect there is a very large quantity of colouring matter: it may reach 50 per cent. of the weight of the dry insect. The colouring matter, which is known as carmine (the same name is also applied to its lake), is soluble in water with a fine red colour. When cochineal is extracted with water the operation must be often repeated, and each time fresh quantities of colouring matter are dissolved. The cochineal may be exhausted by repeated boiling with water, but a large quantity of liquid containing little colouring matter is then produced. In order to obtain a strong solution of the colouring matter the cochineal must be powdered; this is difficult on account of the softness of the material, which does not give a fine powder, but a pasty mass. The object is best attained by grinding the cochineal through a mill similar to a coffee mill, but since the grooves of the steel cone, which effects the crushing, may be easily clogged, the mill must be arranged so that the cone can be taken out and cleaned.

The colouring principle of cochineal is an acid, carminic acid, which was obtained by its discoverer, Warren de la Rue, in the following manner: 1 part of powdered cochineal is boiled with 40 parts of water for 20 minutes; after standing, the liquid is poured off from the sediment, and mixed with a solution of 6 parts of lead acetate, acidified by 1 part of acetic acid. The precipitate, which consists of impure lead carminate, is filtered from the colourless liquid and carefully washed; whilst still wet it is suspended in water, through which sulphuretted hydrogen is passed; the lead carminate is decomposed into lead sulphide and carminic acid, which dissolves in the water. The colouring matter is not yet quite pure; the treatment with lead acetate and sulphuretted hydrogen must be repeated, the solution then obtained is evaporated at a low temperature, the residue dissolved in boiling alcohol, phosphoric acid added to decompose traces of lead carminate still present, then ether, finally the clear liquid is separated from the precipitate and evaporated. By this process carminic acid is obtained in a state of complete purity; unfortunately the process is too complicated for practical application. Pure carminic acid is a purple red mass, which transmits red light at the edges and forms a pure scarlet red powder when ground. The aqueous solution gives, with alum and ammonia, a characteristic precipitate, the colour of which is the purest carmine red; lead, zinc and copper salts produce purple red precipitates. The composition of carminic acid is expressed by the formula $C_{17}H_{18}O_{10}$.

CHAPTER LI.

THE MANUFACTURE OF CARMINE.

THE colouring matter of cochineal is used in the form of a lake under the name of carmine, the finest and most expensive colour used in painting. However simple the preparation of carmine may appear, it is not easy to obtain a product of great beauty; certain conditions, still unexplained, play an important part in the process. Until not long ago the manufacture of fine carmine was regarded as a secret; this it is no longer, and with proper care any one may produce an entirely satisfactory product. Many recipes have been given, some of the best will be mentioned.

Whatever method is used to obtain carmine certain precautions must be taken, without which it is impossible to obtain a bright colour. Alkalis and alkaline earths in very small quantity affect the shade of carmine, so that spring water should never be used in its preparation. Pure rain water, or, still better, distilled water, should be employed.

The decoction of cochineal is difficult to filter. Paper cannot be used, because the pores are so rapidly stopped up that new filters would be continually required. Fine silk is the most serviceable; it should not be washed with soap—the alkalinity of the small quantity of soap the fabric would retain would affect the colour. The decoction is made in a well-tinned copper pan, all other vessels should be of glass or porcelain, which are most easily cleaned, and great cleanliness is the prime essential to the success of the process.

The greatest care is required to prevent the contact of the liquid with iron during the whole process, the smallest trace of this metal would result in a discoloured product.

The essentials of the manufacture of carmine on the large scale are that the colouring matter is dissolved in water, and precipitated by the addition of an aluminium salt, generally alum, absolutely free from iron. The more slowly the carmine separates the finer is its colour. It is generally observed that the last portions to precipitate show the brightest shade. This is because the foreign substances occurring with the carminic acid in the decoction are thrown down with the first portions of the precipitate. In the process of Frau Cenette, famous for the beauty of the product, a solution is made from which the whole of the carmine is separated in about three weeks. During this long time the majority of the substances which have been dissolved in the water together with the carminic acid are decomposed; the liquid acquires an unpleasant smell, and is covered with mould. In the author's opinion so good a product is obtained by this process because the greater part of the impurities is decomposed: the nearer the product approaches the pure compound of carminic acid and alumina, the purer and brighter will be the shade of the carmine.

It is known that light has considerable influence on the beauty of carmine. During dull winter days it is quite impossible to produce so fine a product as in summer. Instead of alum a tin solution may be used to precipitate the colouring matter; the shade of the product is different to that of alumina carmine.

Cenette's Method.—1 kilogramme of finely powdered cochineal is boiled with 75 litres of water for two hours; 90 grammes of saltpetre are added, the liquid boiled for three minutes, then 120 grammes of salt of sorrel (acid potassium oxalate) are added and the liquid again boiled for ten minutes. The

THE MANUFACTURE OF CARMINE.

liquid is then completely clarified by standing, drawn off from the residue by a syphon, and brought into shallow glass dishes which are placed, protected from dust, in a bright light in a uniformly warm place. During several weeks the carmine separates, the last portions being always brighter than the first. The addition of potassium oxalate has the object of assisting the separation of the carmine, for acid salts separate the carmine from solutions; the saltpetre may reasonably be omitted.

The majority of the recipes for carmine, which are frequently sold at a high price, differ but little from the above; acid potassium tartrate is used instead of the oxalate, but the latter is to be preferred because of the slight solubility of the tartrate. It is important not to use too strong decoctions of cochineal, and to add only small quantities of alum. The clear liquid is placed in shallow glass dishes. After a few days the nature of the deposit should be examined: if a considerable quantity of a red precipitate has formed, the liquid is poured off into other dishes, in which carmine again separates during the following days, and usually of a brighter colour than the first. For example, 125 grammes of cochineal are boiled with 5 litres of water during fifteen minutes, 30 grammes of very finely powdered alum are added to the boiling liquid, which is again boiled for a few minutes, allowed to clarify and cool. The greater part of the carmine is then obtained in a few hours, but the liquid still separates carmine after several days.

According to another formula 500 grammes of cochineal are boiled with 30 litres of water, 60 grammes of cream of tartar are added, then 30 grammes of alum, boiling is continued for several minutes, and the liquid then allowed to cool. Carmine may be obtained in a very short time by means of tin solution. The process is similar to that just given. The liquid which would be set aside for the spontaneous deposition of the carmine is, however, returned to the pan, and a solution of pure stannous chloride added in drops, so long as the solution is still clearly red. The carmine separates at the bottom of the pan; the liquid drawn off, even when it appears almost colourless, produces a further small quantity of carmine in a few days.

According to J. J. Hess, a brighter carmine is obtained when the fat of the cochineal is previously extracted by ether or benzene.

Proposals have often been made to deepen the colour of prepared carmine; this must be done with the greatest care, for it is very easy to produce a less handsome instead of a finer pigment. The carmine is moistened with distilled water containing about 5 per cent. of ammonia solution.

Carmine readily dissolves in ammonia. This property may be applied to test its purity, pure carmine should dissolve without residue in 5 to 6 times its quantity of ammonia. Any considerable residue denotes an intentional addition of some adulterant. Starch, vermilion, and cheaper lakes are used for this purpose. The red solution obtained by treating carmine with the above quantity of ammonia may be used as a red ink. This solution can also be used to purify carmine; when it is allowed to stand in an uncorked bottle for some time the ammonia escapes, and the greater part of the carmine is deposited as a very fine powder.

Carmine solution is made by dissolving carmine in the necessary quantity of ammonia, adding glycerine equal in quantity to the ammonia, driving off the latter by heat and diluting the liquid. This solution is well adapted for colouring confectionery, but cannot be employed in painting and writing, since the glycerine would prevent it from drying.

Carmine is extensively used in painting, and for many purposes cannot be replaced by another pigment, *e.g.*, the cosmetic known as vegetable rouge can be made from no other pigment. Carmine may be used in all methods of colouring; it is quite harmless for confectioners' purposes.

Munich, Vienna, Paris, or Florentine Lake.—This fine deep red lake differs from carmine in containing a much larger quantity of alumina, thus possessing the character of an ordinary lake. Occasionally carmine lake is intentionally mixed with light white substances, such as magnesia, to obtain paler shades or cheaper pigments.

Whilst only the finest varieties of cochineal can be used to make fine carmine, the cheaper sorts can be used for Florentine lake. The cochineal used for carmine is boiled but a short time with water, so that the residue contains considerable quantities of colouring matter, which may amount to half of that originally present. The same materials are used as in the preparation of carmine, and in this case, too, stress should be laid upon their purity. The weight of the alum is usually 10 to 15 times that of the cochineal; a little stannous chloride and cream of tartar are also added to brighten the shade. All these materials are boiled with the cochineal, soda is added to the clear solution until effervescence no longer occurs, the separated lake is then washed. The lake may be made with magnesia instead of alumina, it is added in the form of magnesium sulphate; the more magnesia is used the paler is the lake. The proportions in which the materials are employed vary with each manufacturer. The following quantities have always given the author a favourable result: Cochineal 10 parts, alum 150 parts, water 250 parts; or cochineal 10 parts, magnesium sulphate 5 parts, alum 0·5 part.

In making Florentine lake a fairly deep colour should be produced. This may be mixed without difficulty in the wet or the dry state with a white pigment, thus producing any desired shade, even to the most delicate rose red.

Ammonia-Cochineal.—This preparation, which is chiefly used by dyers, is obtained by treating cochineal in a well-closed flask with strong ammonia, which dissolves the colouring matter. After about a month alum equal to about 3 per cent. of the amount of cochineal is added without separating the undissolved residue from the solution, and the whole is evaporated at a gentle heat in a tinned pan until it becomes a stiff paste on cooling. The mass solidifies more readily if a small quantity of starch paste is added. "Cochineal paste" is cochineal which has been treated in this manner; it is generally brought into the market in the form of cakes or small slabs.

In addition to the true cochineal other species of the same insect contain a red colouring matter and have a restricted use in dyeing. The most important of these is the Polish cochineal (*Coccus polonicus*), which lives on the roots of the *scleranthus*. It has not been proved that these insects contain the same colouring matter as true cochineal, but it is certain that the colouring matters of Polish and also of Russian cochineal (*Coccus euvæ ursi*) are far inferior in beauty to that of true cochineal.

CHAPTER LII.

THE COLOURING MATTER OF LAC.

FROM the punctures of the insect *Coccus lacca* on certain East Indian trees, especially those belonging to the genus *Ficus*, flow at the same time resin and colouring matter in such quantities that the insects are frequently enclosed and large red masses are formed on the trees. The insects live in some measure in these masses, the females lay their eggs in the spaces and the larvæ are said to feed on the red sap contained in the mass. When the larvæ have left the incrustation, generally in November, it is broken off from the branches and collected. It is now known as stick-lac; it consists of large masses of resin of a fine deep red colour. In the interior the cells of the insects may still be perceived. On chewing, it becomes soft and colours the saliva a deep violet. When stick-lac is boiled in water a portion of the colouring matter dissolves. Good stick-lac contains 10 per cent. of colouring matter and 80 per cent of resin.

Seed-lac has the same origin as stick-lac, but frequently the greater part of the colouring matter has been extracted; it rarely contains more than 2·5 per cent., and is consequently of little value for colouring purposes.

Lac Dye.—The colouring matter of stick-lac is generally separated from the resin in India and comes into the market under the name of lac dye. It is made by stirring coarsely powdered stick-lac in large vessels for several hours with warm water. Almost the whole of the colouring matter dissolves, whilst the resin remains as a ruby red mass which is melted and brought into the market under the name of shellac. The solution of the colouring matter is evaporated in shallow vessels in the sun or boiled down in pans; the residue is made into cakes. At present soda solution is used instead of pure water to extract the colouring matter; a larger yield is obtained. Lac dye usually contains 45 to 50 per cent. of colouring matter, 25 per cent. of resin, and, in addition, earthy substances which are to be regarded as intentional admixtures.

In the preparation of lac dye by Stephen's method coarsely powdered stick-lac is boiled with soda, and the remainder of the colouring matter contained in the resin extracted by repeated boiling with water. All the extracts are then united and precipitated by alum. The colouring matter separates as a lake which still contains a large quantity of resin.

According to the patented process of Henley, the colouring matter of seed-lac is extracted by hydraulic presses. The lac is filled into press bags which are placed between iron boxes heated by steam. On applying pressure the melted resin, shellac, comes through the bags, whilst the colouring matter remains behind. If this method were employed on the large scale, presses similar to those used in stearin candle works would be suitable.

When lac dye is treated with hydrochloric acid the colouring matter dissolves; the solution can be used to dye wool, which it colours a beautiful red. The colouring matter is very similar in appearance to that of cochineal, but has the appearance of much greater fastness. It appears also to be similar in composition to carminic acid, but little is known as to its chemical constitution.

When lac dye is to be used for painting it must be freed from resin; this is accomplished by treating the finely powdered mass for a long time with boiling alcohol and separating the solution of resin from the undissolved colouring matter, which is then dried. In this way a lake is produced known commercially as Vienna red and little inferior to carmine lake. Unfortunately the price of lac dye, although it is imported in large quantities from India to England, is so high that it can only be used for artists' colours.

CHAPTER LIII.

SAFFLOWER OR CARTHAMINE RED.

Safflower or bastard saffron (*Carthamus tinctoria*) grows in southern Europe, and is also cultivated. It contains two colouring matters, yellow and red. The former is not used, but is employed for colouring liqueurs, since... The red colouring matter is used for colouring artificial flowers and for fine cosmetics; formerly it was used in dyeing, but is now rarely used for this purpose, since carthamine red has little permanence and is replaced by cheaper dyes.

The yellow colouring matter, which is not used, is easily removed by treating the corolla of the safflower, which contains the colouring matter, with water. After this treatment it is known as washed safflower. In the dried corolla of the flower up to 36 per cent. of the yellow colouring matter is found, whilst only 0·4 to 0·6 per cent. of the red is contained.

It appears that the colouring matter of the safflower has been used from the most ancient times. The Chinese employ it to obtain a handsome cosmetic, the Tyrians are said to have used it in dyeing. In Europe it was not cultivated before the seventeenth century, when it came into use for dyeing.

In order to obtain pure carthamine red—and this is the form in which it is generally used—safflower is treated for a long time with water containing a little acetic acid until a yellow solution no longer results. The residue is then treated with soda solution for several hours, when the colouring matter dissolves; the solution is filtered and neutralised with acetic acid. Cotton is then introduced. Carthamine red is a colouring matter which at once dyes animal and vegetable fibres when these are brought in contact with its solution. Thus the whole of the colouring matter is precipitated on the cotton, which is coloured a deep red. After twenty-four hours the cotton is taken out, washed, and treated with soda solution, in which the colouring matter dissolves. When this solution is carefully neutralised with citric acid the colouring matter separates in the form of fine flocks. These are collected, dissolved in strong alcohol, the solution evaporated to a small volume, the colouring matter again precipitated by the addition of a large quantity of water, and washed with pure water until the wash waters begin to be coloured red. The pure carthamine thus obtained is spread upon small cups, in which it dries to a beautiful red mass which in somewhat thicker layers has a fine green lustre. It is brought into commerce as " cup red " or " plate red ".

The small quantity of colouring matter contained in the safflower, as well as the complicated process for obtaining it, make it evident that this colouring matter will be among the most expensive found in commerce. The high price is, however, neutralised by its great colouring power, and by the fact that nothing can in all cases replace it for colouring the best artificial flowers.

Safflower Carmine.—This substance, used by dyers, is a solution of carthamine red in soda solution. It is only necessary to add an acid after once introducing the fabric in order to fix the colouring matter at once upon the fibre.

The shades produced by carthamine red are distinguished

by a delicacy of tint not produced by other dyes; all shades between the deepest red and the palest rose red may be obtained from it. Unfortunately it is very susceptible to the action of alkalis; one careless washing of an article dyed with carthamine is sufficient to remove the greater part of the colour.

On account of its cost and little durability carthamine red is not used in painting. It is principally employed in the artificial flower industry for colouring flowers. Pure carthamine or a mixture with finely powdered steatite is applied by rubbing.

Alkanet.—The roots of alkanet (*Alkanna tinctoria*) contain in the bark a fine red colouring matter, from which also a violet lake can be obtained, though this is rarely made. The plant is largely grown in Southern Europe and also around Vienna. The colouring matter is obtained pure by macerating the root for a long time with water and then treating with strong alcohol, which dissolves the colouring matter, together with a large quantity of resin. The alcohol is distilled off, the residue extracted with ether, and the ethereal extract treated with a large quantity of water, which extracts the colouring matter. This is left in the pure state, when the solution is slowly evaporated.

According to Carnelutti and R. Nasini alkannin is obtained by extracting the root of *Anchusa tinctoria* with petroleum ether, evaporating, treating the residue with weak caustic potash, filtering, and shaking the filtrate several times with ether. The solvent is removed, the colouring matter precipitated by carbonic acid, dried over sulphuric acid, and dissolved in ether. The filtered solution is then allowed to evaporate. Alkannin is then obtained as a dark brownish red mass, which is easily powdered. It has a metallic lustre, and readily dissolves in alcohol, glacial acetic acid and chloroform.

When alum is added to the aqueous solution of the colouring matter a beautiful violet precipitate separates, which when dry is very suitable for artistic purposes. Alkanet is little used. It is completely replaced by much cheaper colouring matters.

CHAPTER LIV.

MADDER AND ITS COLOURING MATTERS.

The root of the madder (*Rubia tinctoria*) contains a red colouring matter which is distinguished over all other colours of vegetable origin by great fastness, on account of which it occupies a most important position in dyeing and colour-printing. The use of madder in dyeing has been much reduced by the discovery of artificial methods of making alizarin, the most important colouring matter of madder. Hence the following account of madder and its products is chiefly of historical interest only.

Madder is cultivated in many countries. The roots, which vary from the thickness of a quill to that of a finger, are cleaned from adherent earth, carefully sorted according to age, and dried. In commerce a distinction is made between stripped and unstripped madder. The cortex of the root contains but little colouring matter and is generally removed by mill stones moving at some distance apart. The stripped roots are finely ground, and are then known commercially as madder. Merchants distinguish a large number of varieties, which are designated according to the place of production. In addition to ground madder there are a number of madder preparations which contain the colouring matter in concentrated form, and are consequently used in smaller quantity. The most important of these preparations are garancin, garanceux, and the madder extracts.

"Refined madder" is made from the natural substance

by subjecting it to a process of fermentation, by which the substances accompanying the colouring matters are largely decomposed, and the product becomes relatively richer in colouring matter.

"Flowers of madder," which have double the colouring power of madder, are made by treatment with very dilute sulphuric acid. Madder is stirred with five or six times its quantity of water to which about 1 per cent. of sulphuric acid has been added. After leaving the mixture in a fairly warm place for five to six days alcoholic fermentation takes place, and many substances which would be injurious in dyeing are decomposed. The residue becomes in consequence richer in colouring matter. When the fermentation is finished the residue is filtered off, subjected to powerful hydraulic pressure, and the mass broken up and very thoroughly dried at a temperature of 60° to 70° C. If it is not thoroughly dried it soon becomes mouldy. This method of treating madder, due to Julians, is simple, and has the advantage that from the alcoholic liquid spirit may be obtained. In some respects this process is more rational than the one immediately following: certain substances which are destroyed in the manufacture of garancin are made useful, and also the process requires little outlay.

Garancin is obtained from madder by treatment with sulphuric acid, which decomposes the glucoside, and also sets free the colouring matter united to lime and magnesia, which would otherwise be lost. In addition, the warm acid radically attacks the nitrogenous compounds, the greater portion of which is destroyed. Thus the residue contains relatively more colouring matter, and the removal of substances which accompany the colouring matter considerably facilitates the dyeing process.

Although the preparation of garancin was recommended in 1828, it was much later before the prejudice which prevented its general employment was completely overcome, and garancin recognised as a very valuable material for the dyer. The simplest process for preparing garancin is as follows: Madder is several times washed with water, then pressed, and the residue mixed with dilute sulphuric acid in a lead-lined vessel. To 100 parts of madder 50 parts of sulphuric acid and 50 parts of water are taken, the mixture is heated by steam to 100° C., and maintained at this temperature for half an hour. By the action of sulphuric acid of this strength considerable charring occurs, and in particular the cell walls of the madder are attacked, so that in the succeeding process the colouring matter is readily dissolved; in consequence of the charring, the garancin acquires a deep brown to black colour. When the action of the acid is finished, it is drawn off and the residue washed with water until it is free from acid, when it is dried. The process here described is the original method; it has been much modified, hydrochloric acid, zinc chloride, alkalis, soap solution, etc., being used in the manufacture of garancin. The colour maker will rarely require to make garancin himself, so that it is sufficient for our purpose to indicate what is understood by garancin, and in what manner it is made.

Garanceux.—Madder which has been once used in dyeing still retains some quantity of colouring matter; this is utilised by treating it by the same process which is used to obtain garancin from fresh madder. The treatment with sulphuric acid almost entirely destroys the cellular structure, so that the residual colouring matter is made accessible to solvents, and consequently the garanceux may be again used for dyeing.

Madder Extract.—The fixation of the colouring matter of madder upon the fibre is attended with many difficulties; attempts were made to present it in the purest form possible, or in such a condition that the operation of dyeing was simplified. The commercial madder extracts consist either of liquid extracts of madder, or of the more or less pure colouring matter itself. Among the numerous solid or liquid madder extracts, without doubt the most important is the crude alizarin (the principal colouring matter of madder), made by the process discovered almost simultaneously by Rochleder and Pernod. This is produced by extracting either madder or garancin with hot water containing sulphuric acid; about 5 grammes of acid are mixed with 1 kilogramme of water and the madder boiled with this dilute acid in a lead-lined vessel. The liquid separated from the solid residue becomes turbid on cooling, and yellow flocks separate which consist of impure alizarin; no further purification is required for practical purposes. Crude alizarin is also obtained by treating madder with superheated steam. These pre parations are important, not only to the dyer and calico printer, but also to the colour maker; from them the madder pigments can be made in a very simple manner.

The Constituents of Madder.—Madder has been most thoroughly investigated: its colouring matter, alizarin, is now made artificially. In addition to woody fibre, madder contains sugar, mucilage, resin, a glucoside (*i.e.*, a substance which can be decomposed into a sugar and another substance), and, in particular, two colouring matters known as alizarin and purpurin.

Alizarin occurs ready formed in madder, and is also produced by the decomposition of the madder glucoside into alizarin and sugar. To obtain pure alizarin finely ground madder is extracted with boiling water; on the addition of sulphuric acid to the decoction alizarin separates together with other substances. The moist precipitate is boiled with a solution of aluminium chloride, and, after filtering, hydrochloric acid is added to the filtrate, when deep red flocks separate, consisting of a mixture of alizarin and purpurin.

The colouring matters are further purified by dissolving in alcohol and adding freshly precipitated alumina, which unites with them; the alumina compounds are then boiled with strong soda solution, in which the purpurin dissolves. The residue consists of aluminium alizarinate mixed with resin, the latter is extracted with ether or benzene, and the residue then decomposed with hydrochloric acid, when alizarin is set free and is purified by recrystallisation.

Alizarin has the formula $C_{14}H_6O_2(OH)_2$. It forms fine red crystals which are very little soluble in cold but more easily in hot water; they dissolve readily in strong alcohol to a yellow solution. Alizarin dissolves in alkaline liquids, the solution is dichroic; it appears dark purple by transmitted light and pure blue by reflected light. When the alkaline solution of alizarin is brought in contact with freshly precipitated alumina the colouring matter is completely thrown down and a beautiful red lake is formed.

The second madder colouring matter is purpurin. This is more soluble in water than alizarin, it also dissolves in alum solutions; its alkaline solutions are not dichroic. The formula of purpurin is $C_{14}H_5O_2(OH)_3$. In addition to these two colouring matters, madder contains a third, known as rubiacin, which also forms red compounds. It appears to be fixed upon fabrics dyed with madder in company with the other colouring matters.

Dyeing and printing with the colouring matters of madder are among the most difficult processes in dyeing; a long series of operations is required to fix the colour permanently on the fabric. The beautiful deep red colour known as Turkey red is produced by madder; it is distinguished from other vegetable colours by its great fastness. By means of madder various other shades besides pure red may be produced. This very valuable substance is largely used in dyeing and calico printing to produce permanent colours.

CHAPTER LV.

MADDER LAKES.

THE madder lakes are equally as permanent as the madder colouring matters dyed upon fabrics. On this account and because of their handsome colour they are highly prized by painters. The best qualities of madder are necessary to produce fine lakes; it is more convenient to use garancin or madder extract. Only when the colouring matter is tolerably free from the foreign substances which accompany it can a pure red madder lake be produced; the poorer qualities of madder produce a lake which is not a pure red and is never bright. The good repute of the madder lakes of certain makers is in great part due to careful choice of the raw materials. This choice and care in the process are the secrets of the manufacturers who produce good madder lakes.

The author has found that the finest lakes can be obtained without difficulty from the alizarin, separated in the before-mentioned manner from madder by means of sulphuric acid. The small cost of obtaining crude alizarin by this process enables it to be used on a commercial scale. The labour involved in the process is far outweighed by the excellent qualities of the lakes produced.

A very bright lake is obtained directly from madder by treating with sulphuric acid (thus garancin is really used), digesting the mass for several hours with a solution of alum absolutely free from iron, filtering and adding at first a small quantity of soda solution; a little madder lake is precipitated which is of the finest shade. When this lake has settled the liquid is poured off and again mixed with a little soda solution. The lake now precipitated is inferior to the first, if the madder is not of specially fine quality. By this fractional precipitation of the colouring matter, madder lakes varying in beauty are produced ; the separate fractions often differ considerably in shade, this is because the later portions are more and more contaminated by the foreign substances contained in the madder.

From inferior madder lake, which could only be sold at a low price, a lake of the best quality can be produced in the following manner : The lake is finely powdered, mixed with acetic acid and left to stand twenty-four hours ; the madder lake dissolves in the acid to a fine red solution, whilst the impurities remain on the filter when the solution is filtered. The clear solution is mixed with a large volume of water free from lime, and the acetic acid neutralised by soda, when pure madder lake separates. The addition of soda is not continued until all the lake is precipitated, but only until the liquid is still slightly red. Hydrochloric acid may be used instead of acetic ; it is much cheaper, but must be absolutely free from iron.

It has already been stated that the crude alizarin prepared as described above is well suited as a source of madder lake, since it is free from the majority of the substances which injure the purity of the shade. It is treated with a boiling solution of alum and filtered whilst hot; madder lake is then precipitated by cautious additions of soda, the first portions precipitated being the best.

Madder Carmine is not often found in commerce ; it consists of the almost pure lakes of alizarin and purpurin. The process by which this valuable pigment is produced is based upon the great stability of the madder colouring matters. The following is the process on the large scale : Good madder, very finely ground, is spread out in small heaps in a room, the temperature of which is about 16° to 18° C.; the heaps are moistened with water and left for several days. The mass ferments and gives off a peculiar smell; in the fermentation not only is the glucoside decomposed, but also many of the other compounds contained in the madder, consequently the mass becomes dark and considerably diminishes in weight. The end of the fermentation process may be recognised by a little practice with tolerable certainty by the disappearance of the peculiar smell; the mass is broken up and transferred to a lead-lined vessel, in which it is mixed with three or four times its weight of ordinary sulphuric acid. The acid is allowed to act for several hours, the plant fibres are almost completely charred, and the mixture becomes black. When the charred residue has settled, the liquid is filtered through pure quartz sand or powdered glass, and mixed with a large quantity of water. The carmine, which is insoluble in water, separates as a red powder; it is then washed and dried.

The colour of madder carmine is of such beauty that it can only be compared with that of good cochineal carmine, but it is incomparably more permanent than the latter and may be used in all varieties of painting.

CHAPTER LVI.

MANJIT (INDIAN MADDER).

THE roots of the East Indian *Rubia mungista* contain purpurin and another characteristic colouring matter, manjistin. This material is a rarity in German commerce, but is imported into England from India in tolerable amount and used in dyeing. Manjistin may be extracted by boiling manjit repeatedly with a solution of aluminium sulphate, uniting the liquids and acidifying strongly with hydrochloric acid. In twenty-four hours a red precipitate forms in the liquid; it is dried and treated with boiling carbon disulphide, which extracts purpurin and manjistin and leaves undissolved a dark resin. The residue, after distilling off the carbon bisulphide, gives up purpurin to dilute acetic acid, manjistin remaining undissolved. When the purpurin has been completely extracted, the residue is treated with a little boiling alcohol and repeatedly recrystallised; manjistin is finally obtained in golden yellow crystals of the composition $C_{14}H_{15}O_2(OH)_2 . CO_2H$.

When a solution of manjistin in a little alcohol is mixed with water no precipitate occurs; on then adding aluminium hydrate and long boiling an orange-red lake is produced. Lead acetate produces a deep orange-red lake when used in just sufficient quantity; with excess a fine scarlet lake is obtained.

If manjit were a common article in the German market, it would be used in the preparation of fine pigments with as much advantage as madder and the various madder preparations.

Chica Red, Curucuru, Carajuru.—The colouring matter known under these names rarely comes into commerce; it forms brownish red masses which acquire a peculiar golden glitter on rubbing. It is obtained from the leaves of *Bignonia chica*, a native of tropical America. The leaves are superficially dried and then covered with water; in the warm climate the macerated leaves soon ferment, and a deep red powder separates at the bottom of the vessel: this is the colouring matter; it is dried, made into small cakes and brought into the market.

Chica red is only partially soluble in strong alcohol, the residue chiefly consists of plant cells. Alum and stannous chloride precipitate beautiful red lakes from the solution, which are distinguished by great permanence when exposed to light. On account of the present high price, this colouring matter has not yet obtained admission into the colour industries, though it would be a valuable recruit.

Chica red has been little studied; an investigation would certainly considerably increase our knowledge of the organic colouring matters. Towards reducing agents it behaves in a similar manner to indigo; in contact with caustic soda, grape sugar and water in closed flasks, a violet liquid is produced; when exposed in an open vessel the violet at once changes to brown, but if the liquid is syphoned off into dilute hydrochloric acid a pure red precipitate separates, which is to be regarded as the pure colouring matter.

CHAPTER LVII.

LICHEN COLOURING MATTERS.

MANY lichens contain compounds from which colouring matters can be obtained by appropriate treatment; the colouring matters known as cudbear, archil and litmus were formerly much more extensively used in dyeing than at present: like so many natural colouring matters they have been replaced by coal-tar dyes.

The lichens which produce colouring matters live upon trees and rocks, chiefly on the sea-shore; the colouring matters are thus generally manufactured in places near the coast. The species principally used for this purpose are *Lecanora, Rocella* and *Usnea.*

The lichens contain colourless compounds, which by treatment with acid or alkalis produce orcinol. In contact with moist air and ammonia, orcinol is decomposed into water and orceïn, which is the colouring matter. The lichens are treated so as to produce the largest quantity of orcinol, which is then converted into orceïn. There is not space here to enumerate and describe the many compounds obtained from lichens, only the most important can be mentioned; these are erythric, lecanoric and usnic acids. Lecanoric acid produces, under certain conditions, orsellic acid, which is again changed to orcinol.

Pure orcinol, $C_7H_8O_2 . H_2O$, is a crystalline substance of sweet taste; it is soluble in ether, alcohol, water, etc. In damp air containing ammonia, orcinol gradually acquires a fine red colour, being converted into orceïn, $C_7H_7NO_3$, according to the following reaction:—

$$C_7H_8O_2 + NH_3 + 3O = C_7H_7NO_3 + 2H_2O.$$

When pure, orceïn is a brownish red amorphous powder, soluble in alcohol to a scarlet solution and in alkalis to a purple solution. On the addition of water to the ammoniacal solution or of salt to the alkaline solution, orceïn is again separated; this procedure may be used to completely purify the colouring matter.

Archil is principally obtained from the lichens *Lecanora* and *Rocella*. These lichens have a wide geographical distribution; archil is made in Sweden, Italy and Spain. It is obtained in a simple manner: the powdered lichen is mixed with decomposing urine to a paste, which is allowed to lie in the air until the colour has changed through red to violet; on the average fourteen days are required for the transformation. More recently the process has been made less disgusting by replacing urine by ammoniacal gas liquor. When the mixture is violet all through the process is finished; the wet mass is packed into barrels, which are well closed to prevent it from drying.

The presence of alkalis accelerates the formation of orceïn: in many districts it is customary to add a little lime to the mixture of lichen and urine; alum is also added to retard putrefaction, which readily occurs in warm climates.

Instead of using lichens to produce the colouring matter, they may be extracted by boiling water, the solution considerably evaporated and then exposed to the air with the addition of ammonia, until the violet colouration appears. Slight additions of sulphuric acid give a more purple tint; soda produces a deeper violet.

French Purple is a compound of the archil colouring matter with lime, *i.e.*, a lime lake. This beautiful substance is obtained by treating the lichen for some minutes with ammonia, pressing out the liquid and neutralising it exactly with hydrochloric acid. The precipitate obtained consists of the lichen acids; it is again dissolved in ammonia and the solution exposed to the air until it has become cherry red. When this is the case the liquid is quickly boiled and placed in large shallow dishes, which are kept at a temperature of 75° C. until the colour is purple-violet. Calcium chloride is then added; a garnet red precipitate is produced, which, when washed and dried, constitutes French purple.

Archil and French purple produce shades of great beauty and purity, but they have little durability and are quickly altered by exposure to light. They are, therefore, seldom used as pigments. The same is the case with cudbear and litmus.

Cudbear is a dirty purple powder with an ammoniacal odour and salt taste. It is made on a large scale in Holland. It differs from archil only in containing the colouring matter in the solid form.

Litmus is generally made from lichens of the genus *Variolaria*, though varieties of *Rocella* and *Lecanora* are often used. It is not made in quite the same manner as archil and cudbear. Together with the ammonia, potash is added to the powdered lichen. The mixture is left until it has become violet, when urine, potash and lime are added at intervals to continue the decomposition until the colour is changed to blue and the whole mass has become a paste. Gypsum and lime are then added, the mixture formed into rectangular plates, and brought into the market, described by a number, according to the amount of gypsum added.

Pure litmus should leave little solid residue when dissolved in water, the solution should be a fine violet. On account of its want of stability this colouring matter is not used in dyeing. It can be used for colouring foods since it is innocuous; but indigo carmine has much greater colouring power and at the same time gives fine shades.

The only use of litmus at the present time is as an indicator to decide whether a solution is acid, neutral or alkaline. In a quite neutral solution litmus is violet. Acids turn the solution to red and alkalis to pure blue. In preparing litmus tincture regard must be given to the fact that the aqueous solution of the colouring matter is always alkaline; acid must be cautiously added until the colour is violet, and the least addition of acid or alkali suffices to change it to red or blue.

CHAPTER LVIII.

RED WOOD LAKES.

RED wood is obtained from species of *Cæsalpinia* growing in South America. The following varieties are distinguished: Fernambuco or Brazil wood, Bahia wood, St. Martha wood, Lima wood and Sapan wood, all of which contain the same colouring matter, but in different quantities. Red wood comes into commerce either as thick logs of a fine red colour or as finely rasped chips. Fernambuco and Sapan woods are accounted the best varieties. Although the colouring matters contained in these woods have been investigated for a long time, an accurate knowledge of them has only recently been obtained. The pure colouring matter of red wood, brasilin, forms orange-coloured needle-shaped crystals, soluble in water, alcohol and ether. On careful heating it partially sublimes undecomposed. Brasilin most probably results from the alteration of a substance contained in the inner less coloured portion of the wood; this compound, brasileïn, has been obtained in colourless needles; its solution acquires a red colour on boiling, and produces the pure colouring matter.

Brasilin is a weak acid of the composition $C_{16}H_{14}O_5$. It forms handsomely coloured compounds with most metallic oxides, which are much used in dyeing. A decoction of red wood gives with alum and soda solution a handsome lake of a pure red colour. With a solution of stannic chloride it at once produces a red precipitate. When a solution of potassium chromate is added to a decoction of redwood, a dark brown precipitate separates, which is a compound of chromium oxide with the partially altered colouring matter. The decoction produces a brownish red compound with ferric oxide.

The decoctions of red wood contain, in addition to the colouring matter, a number of other substances which would injure the shade of the lake obtained from the decoction. The greater portion of these substances may be separated by allowing the decoction to stand for several days, when a dirty reddish brown mud forms at the bottom of the vessel. This simple method of purification is attended with the disadvantage that, especially in summer, the decoction rapidly becomes mouldy. This may be prevented by the addition of a little carbolic acid; not more than 0·01 per cent. of the quantity of liquid is required. The foreign matters may also be separated by adding milk or a solution of glue to the decoction, but the simpler process of allowing it to stand gives the best result.

From the behaviour of brasilin towards the metallic oxides, as given above, it follows that compounds of different colours can be made from red wood. These are utilised in dyeing, but in colour works red wood is only used for the preparation of red lakes. The lakes are found in commerce under numerous names, the most common of which are Venetian lake, Vienna lake, Florentine lake, Berlin lake, etc. According as the colouring matter is combined with alumina alone or with stannic chloride in addition, pale or deep lakes are obtained. The different shades of red wood lakes found in commerce are obtained by additions of a white pigment. For this purpose levigated chalk, gypsum, lycopodium powder, and other white pigments light in weight are used.

In preparing red wood lakes the decoction should always be purified by the process mentioned above; there is then more certainty of obtaining a bright lake. It is advisable to dilute the decoction considerably with pure water before precipitating the lake; experience has shown that brighter colours are thus obtained. To 100 parts of wood 130 to 150 parts of alum are used. The latter is dissolved alone and mixed with the decoction, the lake is then precipitated by soda solution. As the lake separates the liquid becomes lighter; portions should be taken out of the vessel in order to see when all the colouring matter is thrown down. If too much soda solution is added the lake takes a violet shade.

A particularly fine lake is obtained from Fernambuco wood when the extract is acidified with a weak acid, such as acetic, and stannic chloride used together with alum in precipitating the lake. One hundred parts of rasped wood are boiled with 300 parts of water to which acetic acid has been added; the extract is then boiled with 130 parts of alum until the latter is dissolved. To the filtered liquid 20 to 25 parts of stannic chloride are added whilst stirring well, and the lake is at once precipitated by soda solution.

If the lake from Fernambuco wood does not turn out well, a better product may be obtained by treating the lake with such a quantity of hydrochloric acid that a small portion remains undecomposed. This is filtered off, and the colouring matter reprecipitated in combination with alumina by neutralising with soda solution.

By adding a little Dutch pink more orange shades of the Fernambuco lake are obtained. In this manner any intermediate shade between pure red and pure yellow can be produced. It is important to lay down a standard series of shades, and always to make pigments which exactly correspond with one of them. It is not possible to make a determined shade of red wood lake or any other lake obtained from a decoction by working with accurately weighed quantities. In general the result will be near the desired shade, but only by chance will it be hit exactly. This is because different samples of the dye-wood do not always contain the same quantity of colouring matter. It is indeed possible with practice to estimate approximately from the colour of the decoction the shade of the lake which will be obtained from it; but this can only be done exactly by comparing the shade of a small batch of the lake with the standard shades. Thus in the case of a Fernambuco lake it will be seen by comparison with the standard shades whether the colour of the small batch is too deep, too violet, etc.; then by using more alum or more tin solution the defects can be removed.

CHAPTER LIX.

THE COLOURING MATTERS OF SANDALWOOD AND OTHER DYE-WOODS.

IN addition to the dye-woods in more common use, the tropics furnish a number of woods which contain fine colouring matters, but which are only in restricted use, either on account of their cost or because they are not always to be bought. To these belong the camwood or barwood of Madagascar, the East Indian sandalwood, and other woods which have not yet been used for colouring purposes even in the tropics.

The most commonly used of the rarer dye-woods is sandalwood, obtained from *Pterocarpus santalinus*. It is garnet red in colour, sinks in water, and produces a fine red meal, which rapidly becomes brown in air owing to oxidation of the colouring matter. According to recent researches, sandalwood contains two colouring matters, one obtained from the other by oxidation. Only one of these, santalin, is known with certainty. Sandalwood contains about 16 per cent. of santalin, which when pure forms a crystalline powder, differing from other wood colouring matters in being almost insoluble in water. Up to the present it has only been used in dyeing to obtain handsome and fast red colours. From it very bright lakes may be obtained, which may be used equally as well as the red wood lakes. The aluminium and tin lakes are especially fine. The extract of sandalwood is obtained by boiling very fine chips with water, then adding alcohol and leaving the mixture. When stannic chloride or alum, or a mixture of the two, is added to the solution of the colouring matter a red precipitate is produced which, after washing and drying, furnishes a bright red lake.

CHAPTER LX.

BLUE LAKES.

Indigo.—The colouring matter of the indigo plant is distinguished by great resistance to the action of light, air, and chemical reagents. It is thus often cited as a particular example of a fast and durable colour. It is very largely used in dyeing, and in certain combinations is important as a painters' colour.

Indigo comes into commerce in a great variety of qualities, all imported from the tropics, in which alone the plant grows. The principal varieties are Indian, Java, Egyptian, American and African, but in these merchants make numerous subdivisions which are often to be distinguished from one another only by very small differences. Good indigo, whatever its source, has the following characteristics: It appears in almost cubical lumps, but may also form irregular masses. The lumps must show a small specific gravity, a higher might be caused by intentional additions of sand or earth. It is without taste and smell, and the colour of the finest varieties is not particularly bright, but when rubbed with a hard substance, or even by the pressure of the finger nail, it acquires a peculiar coppery lustre, which is the more intense the purer the indigo, and is thus regarded as a common test of quality. Indigo is completely insoluble in almost all solvents; in hot oils a small quantity dissolves, which again separates on cooling. The following are the principal signs by which the quality of indigo is recognised in commerce: It forms a dull-coloured, loose, light mass of earthy nature, a fresh fracture of which has a completely uniform appearance. It must be so porous that it at once adheres to the tongue. Uniformity of colour, light weight and fine fracture are the most important of the practical tests for good indigo. The composition of indigo is well known; we shall glance rapidly over the most important substances which occur in indigo in varying quantity.

The Constituents of Indigo.—Indigo contains the following compounds: Indigo blue, indigo red, indigo brown, indigo gluten, water and salts. The single important constituent for the purposes of the colour maker is indigo blue, or indigotin.

Pure indigo blue can be obtained from indigo only by a lengthy process. The indigo is boiled in turn with dilute sulphuric acid, caustic potash, and strong alcohol. The gluten dissolves in the acid, the indigo brown in the potash solution, and the indigo red in the alcohol. By these operations indigo blue is finally obtained in a state of almost complete purity, it now contains only a small quantity of inorganic substances. The quantity of pure indigo blue obtained in this manner varies, according to the purity of the indigo employed, from 20 to 75 per cent. Indigo blue can also be obtained by direct sublimation from indigo, but the yield is very small, since a large proportion is decomposed by dry distillation.

The most important constituent of indigo, indigo blue, can be obtained in quite pure form by a peculiar chemical process. When finely powdered indigo is brought in contact with a reducing agent in the presence of a strong alkali, the indigo blue is converted into indigo white, which forms a soluble compound with the alkali, which compound is stable only so long as it is protected from the action of the air. In contact with air indigo white is at once oxidised, and is reconverted into indigo blue, which separates from the liquid as a soft deep blue powder ; after washing and drying it is quite pure. Whilst sublimed indigo blue forms needles with a splendid coppery lustre, when precipitated from a solution of indigo white by the action of oxygen it is obtained as an amorphous powder, which acquires the coppery lustre only when rubbed.

The process of the reduction of indigo blue by a reducing agent in the presence of a strong base is not only of theoretical interest, but is largely used in practice. Fabrics are generally dyed with indigo by immersing them in a solution of indigo white (known as the indigo vat), and then exposing them to the air. The oxidation of the indigo white then takes place on the fibre, upon which the indigo is deposited in a very finely divided form ; the blue colour is not immediately formed, there is an intermediate green stage.

Indigo is used principally in dyeing. Certain of its compounds are used as pigments; the starting point in the preparation of these is a compound of indigo blue with sulphuric acid. When indigo is treated with sulphuric acid, according to the duration of the reaction and the temperature at which it is performed, either the monosulphonic acid or disulphonic acid is formed. The reactions proceed according to the following equations :—

$$C_{16}H_{10}N_2O_2 + H_2SO_4 = H_2O + C_{16}H_9N_2O_2 \cdot HSO_3$$
$$C_{16}H_{10}N_2O_2 + 2H_2SO_4 = 2H_2O + C_{16}H_8N_2O_2 \cdot (HSO_3)_2$$

When indigo is treated with eight times its weight of strong sulphuric acid, and the solution considerably diluted, indigotin monosulphonic acid separates as a fine blue precipitate, which dissolves with difficulty in water and alcohol, and is insoluble in dilute acids. In order to obtain the disulphonic acid 1 part of indigo is digested with 15 parts of sulphuric acid at 50° to 60° C. for several days. If fuming sulphuric acid be used the solution is effected in a shorter time. The solution is diluted with a large quantity of water, when a little monosulphonic acid separates ; the clear solution is drawn off and a piece of wool immersed in it. The sulphonic acid is soon fixed by the animal fibre ; the wool is washed with water and then treated with a solution of ammonium carbonate, when a deep blue solution of ammonium indigotin disulphonate is produced. This solution is evaporated at a low temperature and treated with strong alcohol, which leaves a residue of the pure disulphonate, which is dissolved in water, the solution precipitated with lead acetate and the lead indigotin disulphonate treated with sulphuretted hydrogen. The colourless solution becomes blue on evaporation ; finally pure indigotin disulphonic acid is left as an amorphous deliquescent mass. The solution became colourless owing to the reducing action of the sulphuretted hydrogen ; when this was driven off by heating the blue colour reappeared.

CHAPTER LXI.

INDIGO CARMINE.

INDIGO carmine is used in the laundry, for blue inks and as an artists' pigment; in chemical composition it is a mixture of the sodium salts of the two indigotin sulphonic acids described in the last chapter. In working on the large scale it is particularly important that the indigo should be finely powdered and completely free from moisture. It is a laborious process to powder indigo in mortars, and loss always occurs through the formation of dust. Thus when large quantities of indigo are to be powdered it is advisable to use rotating cylinders. These are strong casks bound with iron hoops and provided with a well-fitting slide, through which the indigo and a number of iron balls are introduced. When these casks are rotated for a sufficient length of time the indigo is converted into the finest powder without any loss.

Indigo Mills.—Special mills have been designed for powdering indigo. They may be advantageously employed for other materials also. The construction of a well-designed colour mill is represented in Fig. 36, in which indigo may be very finely powdered without loss. It consists of a cast-iron pan, which can be closed by a well-fitting cover. Through the cover passes the axis of the rotating part, which is moved by the bevel cog-wheels and the handle. Above the cog-wheel fastened to the axis is a cross piece carrying heavy balls at the ends, which serve to keep the motion regular. In the interior of the pan four vertical rods are attached to the axis; these are placed at such a distance that they drive before them the heavy balls in the circular depression of the pan. The action of this apparatus is very simple. After the introduction of the indigo through an opening in the cover the axis is put in rotation, the vertical rods attached to the axis roll the balls before them, and thus break up the lumps of indigo. The larger pieces of indigo oppose considerable resistance to the rotation, which should at first be slow. When the lumps have been broken down the speed of rotation can be considerably increased. After the operation has

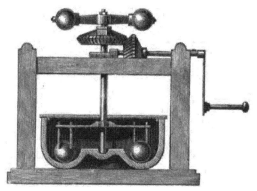

Fig. 36.

been continued for a sufficient length of time the indigo is converted into an impalpable powder.

To facilitate the removal of the ground indigo, an arm carrying a brush of fine hair should be attached to the horizontal arm of the axis. If this brush is placed so that it sweeps the bottom of the circular depression in which the balls move, all the indigo powder in this depression will be carried in one direction. In the depression is an opening closed by a slide during the grinding. The brush expels the ground indigo through this opening when the axis is slowly rotated; it is collected in a vessel below. The larger and heavier are the balls in this mill, the more rapidly will a given quantity of indigo be powdered, and the greater will be the power required to drive the machine. By means of a pulley on the axis of the cog-wheel the mill can be driven by mechanical power.

The powdered indigo is exposed for a long time to a temperature of 120° C., to completely dry it. For this purpose it is spread out on sheets of tin plate in a thin layer and heated so long as a test portion decreases in weight. The powder whilst still warm is at once mixed with the sulphuric acid; 4·5 kilogrammes of fuming sulphuric acid and 1 kilogramme of ordinary sulphuric acid (66° Bé) are used to 1 kilogramme of indigo. Heat is developed in the operation, and in order to prevent the indigo from being charred the vessel in which the mixture is made is placed in a larger vessel filled with cold water. The mixture is stirred with a glass spatula until it is quite uniform. The vessel is then maintained at a temperature of about 50° C. (not more than 60° C.) for 7 to 8 days. After the lapse of this time the indigo is completely dissolved, and the vessel is found to contain two distinct layers, the upper fluid and the lower a paste. The contents are mixed with cold water, about 10 kilogrammes of water to 1 kilogramme of indigo. A solution of 10 kilogrammes of common salt is then added in small quantities at a time. Indigo carmine is insoluble in strong salt solutions; it separates as a deep blue precipitate, which is allowed to settle and then filtered. If it were attempted to wash out the remainder of the salt solution retained by the carmine, a considerable quantity of the latter would dissolve as soon as the salt solution became diluted to a certain extent. To avoid such loss the precipitate is allowed to drain completely on the filter, and then spread out upon bricks, which absorb the water, so that the carmine forms a stiff paste.

Indigo carmine, from which the salt has not been completely washed out, dries completely when exposed to the air. This may be prevented by the addition of glycerine, which is a very hygroscopic substance, and so keeps the carmine always moist, thus preventing the crystallisation of the salt.

Indigo carmine easily dissolves in water to a deep blue liquid. It has tremendous colouring power, in which respect it is only equalled by genuine cochineal carmine. As mentioned above, indigo carmine is extensively employed in the manufacture of ink. The so-called alizarin inks show a very pale colour as they flow from the pen, and only become deep black after some time. To make these inks equally visible at the first indigo carmine is added, by which they acquire a handsome blue colour.

Indigo carmine may also be made by digesting 1 part of dry powdered indigo for 24 to 36 hours with 4 parts of strong sulphuric acid, diluting with water, quickly filtering through a linen cloth, and adding 4 parts of common salt. The precipitate is then collected and dried on gypsum or clay plates.

Blue Lake.—When alum is added to a solution of indigo sulphonic acid, and then soda solution, a blue precipitate is formed, which, when dry, resembles Chinese blue in appearance. It is used in painting. This indigo blue lake is superior to other blue lakes by reason of its fastness to light. It is rather dear, and is mixed with starch, so that it may be sold at low prices; the mixture is formed into slabs and sold as "new blue," "indigo extract," etc. It is chiefly used in the laundry.

Under various names a number of blue pigments are sold consisting of ordinary qualities of indigo mixed with Prussian blue, smalts, chalk, etc. These are used in the laundry, and also for distempering.

CHAPTER LXII.

THE COLOURING MATTER OF LOGWOOD.

THE wood of *Hæmatoxylon campechianum* (logwood), a native of South America, comes into the market in small logs of a red colour; it contains a colouring matter whose properties are fairly well known. Before the discovery of the coal-tar dyes logwood was a most important colouring material; by means of it red, blue, violet and black colours can be obtained. At present it still plays an important part in dyeing.

Many varieties of logwood come into commerce. The most important are Campeachy wood, Honduras, Jamaica and St. Domingo logwood; the first named is the best, and the last the poorest quality. To facilitate the extraction of the colouring matter by water logwood is frequently sold in chips. Many dye-wood grinders moisten logwood with lime-water; this gives the powder a better colour, but diminishes the yield of colouring matter. Since the only useful constituent of logwood can be extracted by treatment with water, and the wood is only a carrier of the colouring matter, which is only useful for fuel when exhausted, logwood extracts are largely used in place of the wood. The extracts are black resin-like masses, which easily dissolve in water; they are very hygroscopic, and should therefore be kept in closed vessels.

Logwood Extract.—Solid extracts can be obtained from the majority of dye-woods. The use of these is a great convenience to the dyer and the colour maker; the method by which extracts are made is therefore briefly described.

Dye-wood extracts can be made on the small scale, by washing out the finely divided wood until soluble matters are no longer taken up by the water, and cautiously evaporating the united extracts. When the extract becomes concentrated the greatest care must be taken to prevent burning on the bottom of the vessel; a burnt extract is always darkened by the presence of products of decomposition, and its solution has a brown colour. Apart from this, the complete extraction of the colouring matter requires much time, and such dilute solutions are produced that they cannot be evaporated, but at the best can be used in the place of water to extract fresh quantities of wood.

When steam is utilised to extract dye-woods these defects are removed. A burnt extract is not to be feared, and a small quantity of water is sufficient to extract the colouring matter completely. The apparatus illustrated in Fig. 37 is very suitable for the extraction of dye-woods, and in general for obtaining vegetable extracts. The extraction vessel is pear-shaped, its two hollow axes move in hollow bearings. Thus it may be turned upside down, and steam and water can be introduced into the interior through the axes, which are connected with the pipes, E, W, and R. The opening at the pointed end of the pear is closed by the screw, S; through it the substances to be extracted are introduced. In the lower portion of the vessel is a sieve upon which the materials are spread out. The opening of the pipe connected with E and W is below the sieve.

The process is commenced by introducing the materials through the upper opening, fastening the cover down steam-tight by the lever, B, and the screw, S, and then running in water through W until it flows out through the lower of the two narrow pipes which are shown somewhat paler in the shaded portion of the drawing. All the taps are then closed with the exception of that on the upper of the two small pipes. By opening the cock on R steam is led in. As a rule, steam at a low pressure is used, not more than half an atmosphere. The contents begin to boil in 15 to 30 minutes, steam then issues from the open tap. According to the nature of the material to be extracted, boiling is continued from 40 to 60 minutes. The side tap is then closed,

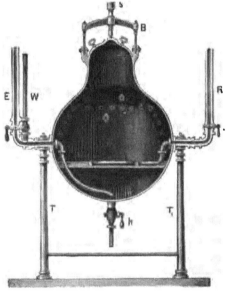

FIG. 87.

and the tap on the pipe E opened. The pressure of the steam now forces the liquid up the pipe E; the pressure of half an atmosphere is sufficient to raise it about 4 metres. In this way it can be forced into a tank above. The rest of the liquid can be run off through the cock, h, at the bottom of the vessel. When quite empty the cover is taken off, the vessel turned over, and the solid residue removed. One extraction of dye-woods is not sufficient to remove the whole of the colouring matter; in most cases the wood is treated a second time before the apparatus is emptied. Even then some quantity of colouring matter remains in the wood. The twice extracted wood is brought into a tub filled with water, and the solution resulting from the long contact of the water with the wood is used in the next operation to extract new quantities of wood instead of fresh water. The dimensions of the extraction vessel vary according to the size of the works. A large apparatus

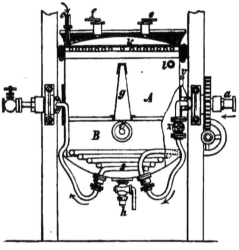

FIG. 38.

costs not much more than a smaller, since in both cases the labour is the same, so that it is advisable to use a large apparatus. When the copper extraction vessel is made to have its greatest diameter about 1 metre, it can be charged with about 50 kilogrammes of rasped wood at once.

The extraction apparatus of Hänig and O. Reinhard is shown in Fig. 38. It can be turned over by means of cogwheels. In the cover are an air valve, e, and safety valve, f; below the cover is a coil, k, fed with cold water from c. In the extraction, by opening the valve v, steam enters through a at l; it is condensed by the cooled cover, the condensed water flows through the material on the sieve, and collects in the space B. After some time v and z are opened, the liquid is then boiled by the steam in the coil, s. Steam rises through the material, is condensed on the cover, and again drops down. At the end of the operation the solution is drawn off through the cock, h.

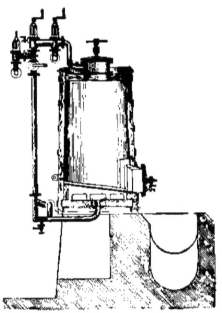

FIG. 39.

Kohlrausch's Process for Obtaining Concentrated Extracts of Colouring Matters and Tannins.—When dye-wood or tan-bark is continually brought in contact with fresh quantities of water, after some time it is exhausted; if the solution containing a certain quantity of the soluble matters is brought in contact with fresh material not yet extracted, it takes up more soluble matter, and thus becomes more concentrated.

THE COLOURING MATTER OF LOGWOOD. 403

The substances contained in tan-bark are very soluble in water, so that by the appropriate treatment of a certain quantity of bark, divided amongst different vessels, with a certain quantity of water, on the one hand the bark can be exhausted completely, and on the other very concentrated solutions ("bark extracts") obtained. The process patented by Kohlrausch is based upon the principles just stated; it can be used to obtain tannin extracts from tan-bark and colour extracts from dye-wood. In this process the raw materials need not be finely ground in order to be completely extracted; they may be used in large pieces. It will be

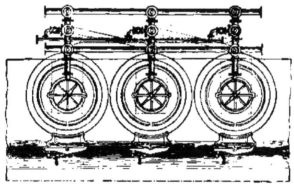

Fig. 40.

understood from the following description that fine bark meal or finely-rasped dye-wood could not be worked.

The apparatus consists of a number of 10 to 20 extractors connected together, also with a water tank above, and with a boiler. A single extractor is represented in section in Fig. 39, together with the necessary pipes and valves by which it is connected with the neighbouring extractors, the water tank and the boiler. Fig. 40 is a plan of three extractors connected together.

The extractors consist of wooden or copper vessels, slightly conical in shape, and of sufficient strength to resist the pressure of one atmosphere. On the top is a copper dome closed by a lid, through which the raw material is introduced, and also steam and water. Immediately above the bottom is an opening, also closed by a screw, which serves to remove the exhausted materials. An inclined sieve is placed at some distance above the bottom; upon this the material rests. The extract collects below the sieve, and may be run off by the pipe or brought into another extractor. All the metallic portions of the apparatus which come in contact with the liquid must be made of a metal such as copper, which does not act upon tannic acid. Iron cannot be used; it forms deep bluish or greenish black compounds with tannic acid, which would cause the extract, instead of being pale and clear, to resemble ordinary writing ink.

In a range of 10 extractors the process is carried out in the following manner: The extractors numbered 1 to 10 are filled with bark or dye-wood and closed; 1 is then filled with water from the tank, and heated by steam to 50° to 70° C. After some time the contents of 1 are forced into 2, and 1 is again filled with water, so that the material in 2 is in contact with the solution from 1, whilst the material in 1 is warmed with a fresh quantity of water under pressure; the extract in 2 is transferred to 3, that in 1 to 2, and 1 is again filled with water, and so on. Finally from 10 a very strong extract of tannin or colouring matter is obtained. The quantity of water required to fill one extractor has come ten times in contact with fresh bark or dye-wood. The material in 1 has been treated with ten times the quantity of water; it is now exhausted, and is replaced by fresh material. The sequence of the vessels is now changed. The original extractor 2 is to be regarded as 1, and 1 as 10. After ten repetitions the original order of the extractors re-obtains. Ninety-nine per cent. of the tannic acid of bark is extracted in this way.

The concentrated extracts obtained in this apparatus

should be mixed with a little carbolic acid to prevent decomposition; they may then be filled into barrels. The extracts may also be so far concentrated by evaporation that they become syrupy. The tannins are readily decomposed; they would be considerably altered if their solutions were evaporated in open vessels. The extracts are therefore evaporated at a very low temperature under diminished pressure in vacuum pans, which are now much used for the concentration of solutions of substances, such as sugar, which would be injured by heating above a certain temperature. Essentially, a vacuum pan is a thick-walled copper vessel, in which the liquid is warmed by a steam coil. It is connected with an air pump, which exhausts the air at the commencement of an operation, and afterwards steam. The liquid is thus constantly evaporated under a low pressure. Extracts of tan-bark and dye-woods boil briskly under these conditions at temperatures below 60° C., at which no decomposition of the tannin or the colouring matter is to be feared. When the solutions have been evaporated to the proper strength they are run off directly into the packages in which they are to be despatched, and in which they become syrupy or even solid masses, according to the extent to which the evaporation has been driven.

The packages should be at once closed; the thick extract is thereby most simply and safely prevented from decomposing, to which risk it would be exposed by the access of mould spores. If these were already present in the barrels or were communicated to the extract by the air, they would either be killed by the hot liquid or would be prevented from developing for a long time. In the closed vessels the extracts remain completely unaltered.

The concentrated solution of colouring matter obtained by extracting dye-woods in the above apparatus can be at once used, after dilution, in the preparation of lakes and in dyeing, but it is not an extract in the ordinary commercial sense of the term, *i.e.*, it does not solidify on cooling. To obtain solid extracts the concentrated solutions must be evaporated; concentration with fire heat would be attended with danger to the quality of the extract, steam heat is therefore used. The liquid to be evaporated is brought into shallow steam-jacketed pans, in which the operation is continued until the liquid solidifies into a resin-like mass when dropped on cold stone. When sufficiently evaporated the extract is allowed to solidify, broken into lumps, and these packed whilst still warm into barrels lined with paper. The lining is necessary on account of the hygroscopic nature of the extracts; when they are exposed to the air they absorb water and form a viscous fluid which soon becomes mouldy.

A properly prepared dye-wood extract should dissolve in water without residue, and the solution when largely diluted should show the characteristic colour of the wood with no brownish shade, and when the colouring matter is precipitated from such a solution by a metallic salt the residual solution should be almost colourless. If the extract dissolves incompletely in water and the solution is brown after precipitation of the colouring matter, the extract has been burnt in the evaporation.

Logwood and logwood extract contain two substances of importance in dyeing and colour making. These are hæmatoxylin and hæmateïn. Hæmatoxylin is found in logwood in greatest amount shortly before the wood is cut. When pure it forms colourless crystals of a peculiar sweet taste, which are soluble in cold water with difficulty, more easily in hot, and readily soluble in alcohol or ether. The composition of hæmatoxylin is expressed by the formula $C_{16}H_{14}O_6$.

Hæmatoxylin is not a colouring matter. It is important

because from it is obtained the essential colouring matter of logwood—hæmateïn. When a trace of ammonia is added to the colourless solution of hæmatoxylin, the liquid at once becomes dark red owing to the formation of hæmateïn. When a larger quantity of ammonia is added the liquid acquires a deep red colour, and then contains only hæmateïn (its ammonia compound), which is formed according to the following equation :—

$$C_{16}H_{14}O_6 + NH_3 + O = C_{16}H_9O_5 . NH_4 + 2H_2O.$$

To obtain hæmateïn in the pure state it is then only necessary to add sufficient acetic acid to decompose the hæmateïn ammonia compound. Hæmateïn separates as dark violet crystals, which readily dissolve in water and alcohol; its aqueous solution gives blue precipitates with the majority of the metallic salts. This behaviour of hæmatoxylin towards ammonia explains the increase in colouring power of logwood, which has been exposed for a long time to the action of the air in the rasped state. Through the action of the ammonia in the air a larger quantity of the hæmateïn ammonia compound has been formed. It has been proposed to facilitate the formation of this ammonia compound by moistening rasped logwood with a very dilute glue solution and allowing it to lie in the air. This process can only have the object of utilising in the formation of hæmateïn the ammonia resulting from the decomposition of the glue, but in this decomposition deep-seated reactions occur, which might affect the hæmateïn itself. It thus appears more suitable to effect the formation of hæmateïn by the direct use of ammonia. This can be done with little cost by watering heaps of rasped logwood with ammonia and repeatedly shovelling about the wood so that it comes into contact with the air. The author has found that the conversion of hæmatoxylin to hæmateïn is very complete in this process; care should be taken not to make the layer of rasped wood too deep, and to take its temperature frequently. In the transformation of hæmatoxylin to hæmateïn the temperature rises, the rise might be injurious if it proceeded too far. Thus if the temperature of the interior of the heap is found to be high the wood should be turned over.

The solution of the logwood colouring matter produces handsome lakes, all of which have, however, the inconvenient property of acquiring an ugly grey colour on long standing. The finest and most durable of the logwood lakes is known as violet lake, which is made with alumina salts. The best result is obtained when a solution of aluminium acetate, obtained by precipitating alum with lead acetate, is mixed with a logwood decoction or a solution of logwood extract. The precipitate is pale or deep violet according to the amount of aluminium salt added. After drying to a certain point at a gentle heat it can be mixed with gum solution to a paste, which is then completely dried.

Logwood is most valuable in dyeing and calico printing, in which it serves to produce a fine and durable black. When potassium chromate is added to a decoction of logwood a deep black liquid results, which can be used as a good and cheap writing ink. If somewhat stronger solutions are used a greenish precipitate first separates, which soon acquires a pure black hue; it is the chromium lake of hæmateïn. This compound is very durable and is largely used in dyeing to produce fast blacks. The black precipitate might be dried and used as an artists' black pigment if the carbon blacks were not cheaper and more durable.

CHAPTER LXIII.

GREEN LAKES.

THE yellow colouring matters produce green compounds with copper salts. These are important because they can be produced without great cost, and are not as poisonous as the arsenic pigments, although copper compounds are far more poisonous substances than should be used for certain purposes. Their employment might be attended with danger if used to colour children's toys and pictures.

These green copper lakes are simply made by adding copper sulphate solution, free from iron, to a hot decoction of yellow berries or weld until the liquid is emerald green, and then adding caustic soda solution in small quantities to precipitate the lake. The temperature of the liquid should not be more than 50° to 60° C. when the caustic soda is added. It has been maintained that these lakes turn out well only when the precipitation is so conducted that the residual liquid is quite colourless. The author has, however, found it more satisfactory to discontinue the addition of caustic soda when the liquid is somewhat coloured, since when the precipitation is complete other substances besides colouring matter may be precipitated.

Chlorophyll.—All the higher plants contain the same green colouring matter—chlorophyll. It has a very fine shade, is fairly fast, and the raw material from which it is produced is to be obtained at nominal cost. Special attention should, therefore, be paid to this colouring matter.

Chlorophyll is very easily obtained. Grass of a good green, green leaves, or any green portion of a plant is allowed to stand with weak caustic soda solution in a large vessel for twenty-four to thirty hours. The liquid is poured off, brought to the boil for a moment, at once filtered and neutralised with hydrochloric acid. The chlorophyll, which has been dissolved by the caustic soda, is then thrown down as a grass green precipitate; after washing and drying it can be used as a pigment. A chlorophyll lake can be obtained by dissolving the colouring matter in caustic soda and adding alum solution. The green precipitate is a compound of chlorophyll and alumina.

Tschirch gives the following method for obtaining chlorophyll: Green leaves (grass), not containing tannin, are extracted with boiling alcohol, the extract cooled, filtered and evaporated until a sticky mass remains. This is washed with hot water until the washings are colourless, and the residue extracted with cold alcohol. The solution is evaporated to half its volume, the separated crystals are dissolved in alcohol, and the solution heated on the water bath with zinc dust. A deep emerald green solution with a red fluorescence is obtained; it can be preserved unaltered for a long time in blue glass bottles. It retains its colour in diffused daylight also for a lengthy period. A beautiful green chlorophyll lake is obtained by boiling this solution with the solution of an aluminium salt and precipitating with soda. Unfortunately, the lake is not fast. The chlorophyll solution obtained by this process is quite innocuous, and is well adapted for colouring liqueurs and confectionery.

Sap Green.—This lake, also known as bladder green, is

obtained from unripe yellow (Persian) berries. It can be used as a water colour, but not in oil painting. This pigment differs from the ordinary yellow berry lake in colour and chemical composition; the yellow lake is a compound of xanthorhamnin with the metallic oxide, whilst sap green contains an uncrystallisable bitter substance, rhamnocathartin.

The lake is obtained from yellow berries which are not quite ripe. They are broken up and left in a warm place. The mass soon ferments; after about ten days it is pressed. Four parts of the liquid are then mixed with 0·5 part of alum and 0·5 part of potassium carbonate. The salts are dissolved in boiling water, and the solutions added to the hot sap. The mixture is then evaporated to the consistency of a thick syrup, which is generally packed in animal bladders, hence the name of "bladder green". Some care is necessary in the evaporation to prevent the burning of the soft mass on the bottom of the vessel. If the evaporation is carried somewhat further the mass solidifies on cooling; it is then black, and only transmits green light at the edges. The use of potash in preparing sap green is attended with the disadvantage that the hygroscopic nature of the salt prevents the colour from drying. If magnesia is used instead of potash the lake dries far more rapidly, but has much less covering power. Sap green is principally used for colouring paper and leather.

Orange-coloured lakes also can be obtained from yellow berries by precipitating the decoction with stannic chloride. These lakes are not directly employed, but are produced immediately upon the fabric in dyeing.

Chinese Green, Lokao.—Under this description a green lake has recently been imported from China. It is a valuable pigment. Chinese green is sold in flat cakes, which are blue, with a green or violet lustre. The powdered lake is pure green and partially soluble in water; it contains colouring matter, water and inorganic matter consisting chiefly of clay and lime.

Chinese green is made in China in a peculiar manner from the twigs of certain species of *Rhamnus*. According to report, the bark is boiled with water and a cotton cloth immersed in the decoction. The fibre fixes a colourless substance which becomes green on exposure to air. The cotton is repeatedly dipped into the decoction until it has absorbed a large quantity of colouring matter. It is then washed with cold water and boiled with water upon which cotton yarn is laid. The colouring matter suspended in the liquid adheres to the yarn, which is washed with a little cold water, and the colouring matter then collected upon paper and dried.

If Chinese green is really made by this process, it is not a lake but a vegetable colouring matter, to which clay has been added to increase its weight or to make it plastic. Very fine lakes can be obtained from Chinese green by dissolving it in a solution of alum, and adding soda solution. When Chinese green is dissolved in acetic acid and ammonium chloride and a zinc salt added, a blue lake is produced on the further addition of sodium acetate. A bluish violet lake is obtained by treating Chinese green with a strong reducing agent and adding calcium acetate.

Charvin's Green.—The very high price at which genuine Chinese green was sold whilst it was still a novelty, gave rise to attempts at imitation. Charvin, of Lyons, succeeded in producing a pigment from *Rhamnus* which has the characteristic property of Chinese green of retaining its colour in artificial light. He plunged the bark of common buckthorn into boiling water, boiled for a few minutes, and allowed the whole to stand for twenty-four hours. Limewater was added to the brown liquid, which was then exposed to the action of the air in shallow vessels, when it gradually turned green and deposited a green precipitate. When this appeared, the liquid was brought into glass vessels and potassium carbonate solution added so long as a precipitate resulted, which after drying had all the properties of genuine Chinese green.

Although for some time Chinese green was a very fashionable colour, its use is now almost discontinued, although it can be used with advantage as an artists' pigment, and its preparation by Charvin's method is neither difficult nor specially costly.

CHAPTER LXIV.

BROWN ORGANIC PIGMENTS.

Asphaltum is found in large deposits in several regions of the earth. It is a compound of carbon and hydrogen, and varies in appearance from cobblers' wax to tar. Asphaltum is of organic origin and is closely related to the mineral oils. Among the pigments usually employed by artists asphaltum plays an important part; it produces very warm shades between brown and deep black. The preparation of asphaltum for artists' use is very simple. Good uniform lumps, free from sand and other impurities, are coarsely powdered and mixed with a solvent in a well-stoppered flask. Asphaltum readily dissolves in essential oils, and also, though with more difficulty, in fatty oils. It dissolves slowly at the ordinary temperature, but quickly on warming. On account of the inflammability of the essential oils generally used, certain precautions must be taken. The asphaltum is mixed with turpentine in a large flask which is well closed and heated by hot water. In a short time a viscous mass is formed which is mixed with the solvent by shaking.

Asphaltum may also be prepared, without dissolving, by powdering and grinding with oil exactly as a mineral pigment.

Sepia.—This brown pigment has a warm hue which is not readily surpassed. It is an animal product; the cuttlefish, a species of cephalopod which abounds in all warm seas, produces a peculiar colouring matter which is stored in a sac, commonly called the ink-bag. It is used by the animal for protection, when pursued it ejects the contents of the bag, and thus makes the surrounding water so dark that it is enabled to escape.

Sepia is made almost exclusively in Italy. The contents of the ink-bag are quickly dried, and then rubbed with strong caustic lye to a thick pulp. More caustic lye is then added, and the mixture heated almost to boiling; from the filtered solution sulphuric acid precipitates the pure colouring matter.

As one of the most handsome brown pigments, sepia is largely used, but unfortunately it can only be used as a water colour. On account of its cost, sepia is frequently imitated. Vegetable substances are charred, extracted, and the extract concentrated until it solidifies on cooling; it is then finely powdered, made into a paste with gum Arabic or tragacanth, and formed into cakes. All such imitations are, however, so imperfect that they are at once recognised. On comparison with genuine sepia, none of them is found to possess the warm shade peculiar to sepia.

Brown colouring matters can be obtained by heating all soft portions of plants. The products have a deep brown shade in consequence of the high proportion of carbon they contain. When the young twigs of soft woods are exposed in closed cylinders to a temperature of about 300° to 400° C., and the residue powdered, colours ranging from a dark rust brown to almost pure black are produced. The higher the temperature employed the more nearly the shade approaches to black.

Parts of plants which contain sugar or similar compounds become deep brown at comparatively a low temperature. The colour, which varies from brown to hyacinth red, is due to caramel. This substance is produced when coffee, beetroot, or chicory root, all of which contain a large proportion of sugar, are heated. Such pigments are little used; they cannot be ground with oil, and in water can rarely be mixed with other colours.

CHAPTER LXV.

SAP COLOURS.

UNDER the designation of "sap colours" several pigments are brought into the market in the condition in which they are ready for immediate use in printing. The term is practically restricted to lake pigments which form a transparent layer when dry, thus the sap colours may be defined as dissolved lakes mixed in a viscous medium, such as thick gum solution. In general the sap colours are not much used, since when wetted with water they again dissolve, which is not the case with pigments ground in oil. Yet for certain purposes they are commonly used, as in the manufacture of playing cards.

In the manufacture of these pigments only colouring matters soluble in water can be used, the number of which is restricted. According to a particular method a tin lake is first made, which is decomposed by a strong base, such as ammonia, so that the colouring matter again goes into solution. The very deep coloured liquid is then mixed with a thickener and an indifferent white substance such as flour or starch, formed into thin sticks, and sent into the market.

A decoction of buckthorn berries is used for the yellow sap colours. It is considerably evaporated, and then mixed with 2 to 3 per cent. of alum, after which starch paste or gum Arabic and sugar are added in proper quantity, and the mixture evaporated at a low temperature so that it is not browned by over-heating.

When a tin lake is employed in the preparation of a sap colour, the precipitate obtained by adding stannic chloride to the solution of the colouring matter is washed, and, without drying, treated with a small quantity of strong ammonia; generally about 10 per cent. of the volume of the precipitate is sufficient to dissolve the lake. It is always better to leave a portion of the precipitate undissolved than to modify the colour by the addition of too much ammonia. The precipitate should be dissolved in a glass vessel; the liquid is well stirred, covered over, and allowed to stand until the undissolved portion of the precipitate has settled. The ammoniacal solution of the colouring matter cannot be concentrated by evaporation. Thus, before the precipitate is dissolved it must be well freed from water by long draining. To give the solution the proper consistency about an equal quantity of thick gum solution is added, and then sufficient starch to form a paste which can be rolled into thin sticks, which are then dried upon boards at a gentle heat. Sap colours are made in exactly the same way from any other colouring matter adapted to the purpose. In using these colours it is simply necessary to bring them into water; if they contain only gum and sugar in addition to colouring matter they completely dissolve, but if they contain starch the solution is incomplete. As a rule starch is only added to give the colour the consistency requisite for printing.

Red sap colours can be made either from red wood lake or from cochineal carmine; the former is used in ammoniacal solution, the latter as a solution of the pure carmine in ammonia. The colour is prepared from red wood by allowing a decoction to stand for several days, adding 2 to 3 per cent. of alum, evaporating and thickening with gum solution. This colour is cheap, but cannot compare in beauty with that obtained by treating red wood tin lake with ammonia. The solution of genuine carmine in ammonia produces a sap colour which leaves nothing to be desired in regard to shade, but on account of the cost of carmine it can but rarely be used. A solution of pure indigo carmine mixed with gum solution is used as a blue sap colour; it can be shaded by additions of red or yellow colours. Chinese blue may be used instead of indigo carmine; it dissolves in a solution of oxalic acid. When Chinese blue, which has been thoroughly washed, is mixed whilst still moist with a little saturated oxalic acid solution, the blue dissolves, and the solution may then be made into a paste by the addition of thickening materials.

Green sap colours are made, as a rule, by mixing yellow and blue. All shades can be thus obtained. A green colour of different composition is obtained by boiling the violet solution of chrome alum with gum solution until the colour changes to green. Violet colours are obtained by mixing red and blue. Sepia produces a brown sap colour.

It has been proposed to use glucose in place of the much dearer gum Arabic, and also to use malt extract as thickener. These substances absorb water from the air, and thus the colours prepared with them never become quite hard, but always remain pasty. For some purposes this is a decided advantage, since such colours can be readily taken up by the brush and easily rubbed up with water, but they cannot be formed into cakes as is usually done with water colours. When the colours are to be made into cakes they must be thickened with gum Arabic alone, without glucose or malt extract. The cakes must be dried with great care; if they are dried too quickly they will crack or even fall to pieces.

CHAPTER LXVI.

WATER COLOURS.

THE colours thus called are so prepared that they easily mix with water to such a condition that they can be applied with the brush. Special pigments are not required for this purpose, ordinary dry colours are simply mixed with a binding medium soluble in water, and the paste is generally pressed into cakes. In commerce very different qualities of water colours are found, ranging from cheap children's playthings to the most costly colours used by painters in water colour. Those pigments which are not already in very fine powder, in consequence of the process by which they are made, such as chrome yellow, must be subjected to a very careful process of levigation. To simplify as far as possible this laborious process, it should be conducted by grinding the pigment as finely as possible, well stirring it in a tub with water, allowing to rest for several minutes, and then running off the liquid into a second tub, from which it is again drawn off into a third after several minutes. The liquid remains in the last vessel until it is quite clear, when the deposit is collected. This is now so fine a powder that a powerful microscope is required to distinguish the separate particles. The residue in the first and second vessels consists of the coarser particles; it is again ground with fresh material. In this process regard must be given to the specific gravity of the material: the higher it is the shorter is the time during which the liquid is allowed to remain in the tubs, since the coarser particles of heavy substances settle very rapidly. In dealing with the light lakes, such as the alumina lakes, the liquid in which the lake is suspended must remain much longer at rest. Two vessels may be used instead of three, without danger of the precipitate containing coarse particles.

Gum Arabic and tragacanth are used as binding materials for the pigments. Dextrine is also much used in place of the expensive gums; for this purpose only pure white dextrine should be used, since the brown colour of ordinary dextrine would injure the shade of the colours, especially of pale colours. Gum Arabic and dextrine, which are readily soluble in water, require no special preparation before their solutions are mixed with the pigments. The solutions are made by treating with water; they are then allowed to stand for several days in a tall vessel so that impurities may settle. If the solutions are very turbid, they are filtered through closely woven linen.

Gum tragacanth requires rather different treatment. It is not completely soluble in water, in which it only swells up to a great extent. It is prepared by leaving it for several days in water, and when it has swollen, rubbing the slimy mass in a mortar until it is completely uniform.

The levigated pigments are allowed to dry in the air to a soft paste, which is mixed with the proper quantity of gum Arabic and tragacanth solutions. As a rule the two gums are used together. The colouring matter and binding medium may simply be ground together, but long grinding would be required to produce a completely homogeneous mixture. The costly manual labour is therefore as far as possible replaced by machinery, by which means a cheaper and also more uniform product is obtained. The machines by which the pigment is ground with the binding material are of simple construction. They consist of rollers placed in pairs one above the other and moving in opposite directions. The two rollers of each pair are connected by cog-wheels, in such a way that the lower moves more slowly than the upper. In consequence of this arrangement, the rollers, in addition to crushing, exert a grinding action upon the viscous mass passing between them. As a rule, after the colour has passed through one pair of rollers, it goes through a second and third, from the last of which it is taken off by a scraper. It is now completely uniform.

The binding medium must be of such consistency that, after grinding with the pigment, a fairly stiff paste results which is suitable for pressing. The cakes of colour are pressed out in an ordinary spindle press, which should be so constructed that the stamp in the down stroke comes upon the paste beneath it and stamps out cakes upon which the engraving of the die is clearly shown. If the cakes crack when slowly dried, the medium contains too much gum Arabic. If the impression of the die upon the cakes is not sharp, and if they remain rather elastic when completely dry, too much tragacanth has been used. It is rather difficult to grind a stiff paste, many manufacturers prefer to grind the colour in a rather more fluid condition, and then to allow the paste to thicken by drying to the consistency necessary for the production of good cakes.

The moulds in which the cakes are made must be very carefully worked in metal, so that the description and trade mark print clearly upon the cakes. The cakes should be dried at the ordinary temperature or but little higher. They are placed upon smooth boards, and care is taken that the temperature of the drying-room remains uniform, which is the condition requisite for the production of the fewest cracks. When it is intended to place on the market only faultless and well-stamped cakes they must be sorted when dry, and the cracked ones rejected; these can be worked up in the next operation.

The dry cakes are given a good appearance by coating them with a weak solution of gum, and then drying. According to the price at which these colours are to be sold the cakes are given a different character. The finest colours are generally made into larger cakes and packed in handsome boxes, whilst ordinary cheap colours are made into small lumps or circular plates, flat on one side and somewhat convex on the other, and packed in boxes of soft wood.

Moist Water Colours.—Instead of grinding water colours with gum Arabic or tragacanth and bringing them on the market in the dry state, they may also be sold in a condition resembling that of oil paints. This may be accomplished by using very viscid glucose syrup instead of gum Arabic, and grinding the pigments in this exactly as in oil. Glucose is very hygroscopic, the colours prepared with it remain moist, and may be spread out upon the palette like oil colours. It is then only necessary to wet the brush in water and mix the mass with it in order to obtain colour of the proper consistency.

Moist water colours have also been known as "honey colours," since this mixture of sugars was formerly used for their preparation; it also is hygroscopic. Honey is no longer used; the much cheaper glucose answers the same purpose. On account of their semi-fluid nature, moist water colours are put up in tubes just as oil colours. They are thus rather dear, but are little used.

CHAPTER LXVII.

CRAYONS.

CRAYONS are coloured pencils by which pictures may be, so to speak, "dry painted". At present this method of painting is little used, though it was in vogue in the last century, but coloured pencils, especially blue and red, are much used for writing. Crayons are now used in a very convenient form, being generally made in the same manner as lead pencils. The coloured mixture from which the crayon is made is produced in the form of a paste containing a soft, finely-ground mineral, well mixed with the colouring matter and an amount of binding medium just sufficient to hold the powders together.

Gypsum is generally used as the soft white mineral forming the base of the coloured mass. It is much better to use steatite (soap-stone), which is not much dearer, and has many advantages over gypsum. The preference given to steatite is based upon a comparison of the properties of the two minerals; gypsum is crystalline, steatite is non-crystalline. Powdered gypsum has essentially a dry character, whilst powdered soap-stone is of a peculiar greasy nature, and consequently can be readily smeared upon a flat surface; it also imparts a pleasing lustre to the colours with which it is mixed.

The manufacture of crayons consists in preparing and moulding the coloured mass. The process is commenced by mixing the powdered colour and soap-stone in a closed rotating cylinder. The rotation is continued until the mixture is uniform in colour. The quantity of colour to be mixed with the soap-stone depends upon the shade to be produced. Manufacturers who make a speciality of crayons prepare a number of shades of each colour. The artists who use crayons, however, usually require only those of a pure shade, since they themselves can produce the intermediate shades from the essential colours. On this account it is advisable to make crayons of distinct and deep colours. The mineral pigments are the most suitable : for yellow, deep chrome yellow ; for red, vermilion or deep madder lake ; for green, one of the pure green pigments, such as chrome green ; for blue, Chinese blue or ultramarine ; for brown, burnt sienna or manganese bistre. For white and black crayons levigated chalk is used, alone or mixed with a sufficient quantity of fine vine black or some other good black. Either gum Arabic or size may be used as binding material. When the former is employed the crayon becomes so brittle when dry that it breaks when pointed with the knife, in spite of the greatest care. Size produces a less brittle crayon, and is also much cheaper, so that it is preferred.

The paste from which the crayons are to be formed is made by mixing thin size with the colouring matter and steatite to a soft pulp, which is then kneaded to secure uniformity. The process then proceeds in a different manner according as the crayon is to be pressed or sawn out.

For producing the crayons by pressing, a simple apparatus is required. It consists of a horizontal metal cylinder with a well-fitting piston. The front of the cylinder is closed by a metal plate in which is an orifice of rather greater diameter than the crayons to be made. In front of the cylinder is an endless band which moves away from it. In using this apparatus to shape the crayons, the paste is made of such consistency that a slight pressure forces it out of the orifice in the front of the cylinder in a coherent rod. When the cylinder is filled with the paste, care must be taken that it includes no air bubbles, for these would cause the rod to break. The piston is then put in position and the mass forced by uniform gentle pressure out of the narrow opening. The endless band must move away at the same rate as the rod proceeds from the cylinder, so that a long stick of the crayon mass rests upon it. This stick is cut up by a blunt knife into uniform lengths, which are dried upon boards covered with blotting paper, and are then enclosed in casings similar to those of the ordinary lead pencil. The rods must be given a rather greater diameter than they are to possess when dry, since they shrink somewhat in drying.

The second method consists in producing a thick paste which is moulded into blocks of the length of the crayons. These blocks are very slowly dried at a uniform temperature, and when completely dry, are cut by a fine saw into thin rods, which are then enclosed in a wooden case. The powder produced in sawing is used in the next operation. This simple process has many drawbacks. In the first place it is difficult in drying large blocks of the crayon mass to avoid cracks, which must be carefully filled with thin paste. Also in sawing up the blocks a large number of rods will be broken, even with the most careful treatment (these are again worked up with the powder produced in sawing). Thus, generally, the formation of crayons by pressure is preferred.

It is important to dry the crayon rods so far that they do not shrink further after they are placed in the case; otherwise they would break up when sharpened. At times it is quite impossible to make a usable point on such a faulty crayon.

Crayons for Earthenware.—Crayons for this purpose are

made, according to M. Rösler, by finely powdering colours suitable for glass and earthenware, and mixing them to a paste with a solution of 2 parts of gum Arabic and 1 part of Marseilles soap. Pencils are then moulded from the paste and wrapped in a strip of paper just previously dipped in a paste of gypsum. Upon unglazed articles or ground glass these crayons can be used like a lead pencil. The colour is fixed by burning in the usual manner.

CHAPTER LXVIII.

CONFECTIONERY COLOURS.

CONFECTIONERS employ several colours. Yellow, brown and black are usually obtained by means of caramel; the other colours, red, green and blue, which are also largely used, must be made from harmless compounds. Unfortunately, colour makers have been known to offer the consumer for this purpose colours which by no means correspond to this requirement; even the poisonous arsenic pigments have been used for colouring sweets. In such a case the user of the colours is less to be blamed than the manufacturer, who should sell for this purpose only colours in no way injurious to health. Fortunately we possess among the pigments of organic origin a sufficient number to satisfy these requirements: for green, either sap green or leaf green may be used without hesitation; for red, cochineal carmine is well suited, and for blue, indigo carmine. The two pigments last named are indeed expensive, but since they possess great colouring power they can still be used for colouring cheap sweets.

For colouring liqueurs equal care must be taken to choose non-injurious colours. Those mentioned above may also be used for this purpose. Recently these colours have been largely displaced by the aniline dyes, which are particularly adapted by their beauty and great colouring power for the colouring of confectionery or liqueurs. There should be some hesitation in using these dyes for colouring articles of food. Some of them are made by means of arsenic compounds, and it is very difficult to free them absolutely from every trace of arsenic. But supposing the dye to be free from arsenic, there may still be objections to its use, for the pure dyes themselves may possess poisonous properties, and thus ought not to be used for colouring articles of food.

For colouring foods the colours above mentioned are sufficient. Orange is obtained by mixing caramel with red, violet by mixing red and blue, and the shades thus produced are quite sufficient for the purpose in question. The colour maker should sell the colours in such a condition that they can be used without further preparation. The colours for confectioners and liqueur makers should be put on the market in a semi-fluid or pasty condition. For this purpose carmine should be ground with a very thick sugar syrup. Leaf green and indigo carmine require no further preparation; they are already semi-fluid, they readily diffuse, and are also soluble in alcoholic liquids.

CHAPTER LXIX.

THE PREPARATION OF PIGMENTS FOR PAINTING.

ACCORDING to the purpose for which they are to be used pigments require a different method of preparation. The preparation of pigments for certain special purposes, such as sap colours, cake colours and water colours, has been already described. The present chapter deals with the preparation of those pigments which are used in large quantities by artists and for ordinary painting.

Colours for artistic purposes require different treatment to those used for ordinary painting. It is important that both should be ground to a completely homogeneous mixture with the binding medium. At first sight this appears to be a simple operation, but in practice there are difficulties which are not too easily overcome. Artists' colours are generally ground with a drying oil. The drying oils are vegetable oils which, when exposed to the air in a thin layer, in a short time become very viscous, and finally completely resinify; linseed, poppy, and nut oils possess this property. As a rule, artists' oil paints are ground with poppy oil. To prevent the paint from drying to a solid mass by exposure to air it is enclosed in protecting vessels. Formerly the ground paint was sold in small bags of bladder. These are now no longer used; instead, collapsible tubes made of a soft tin alloy are employed. They are closed at one end; the other has a neck upon which a metal cap screws down. Pigments ground with just sufficient oil to form a thick paste may be preserved in these air-tight tubes without altering in consistency.

Cheaper pigments, such as white lead, white zinc, chrome yellow, etc., which are principally used in ordinary painting, are ground with raw or boiled linseed oil. Ordinary boiled oil is made by boiling linseed oil with litharge; it contains a certain quantity of lead in solution. It has been repeatedly stated in this work that lead compounds are very sensitive to sulphuretted hydrogen. Pigments which are not themselves altered by sulphuretted hydrogen acquire a dark shade when they are ground with ordinary lead-boiled oil, because the lead is slowly but certainly converted into black lead sulphide on exposure to air. Thus, in order to retain the original beauty of colour, lead-boiled oil should be replaced by oil boiled with manganese borate, which is at least as cheap and has the advantage that the oil does not darken in air. Now that zinc oxide is so cheap it is more and more used in place of white lead, it is not sensitive to sulphuretted hydrogen, and does not even change its colour in an atmosphere of the pure gas. It seems to be quite illogical to use this pigment with a lead-boiled oil. A coating would be quite discoloured in the course of time, whilst, if used with manganese-boiled oil, it would retain its white shade unaltered.

The dry pigment was formerly mixed with the oil by manual labour. The pigment was spread out upon a smooth stone slab, the oil poured over, and the two substances rubbed to a uniform paste by means of a glass or stone rubber known as a muller.

Paint Mills are constructed in various ways. The mixture of colour and oil is ground either between two metal plates or between two rollers pressed close together.

THE PREPARATION OF PIGMENTS FOR PAINTING. 431

Fig. 41 represents the construction of a paint mill in which the grinding is accomplished by a rotating disc. The mixture of pigment and oil stirred together is brought into the hopper, T, from which it must pass between the rotating grinding disc, M, and the lower surface of the hopper. The two substances are thus mixed together. The ground paint flows out of the ring-shaped vessel surrounding M into a receiver below. The grinding disc is driven by means of horizontal and vertical toothed wheels, the latter of which is

FIG. 41.

connected with a pulley driven by power. By means of a screw on the lower part of the plate in which is the bearing of the axle of the grinding disc, the distance of the latter from the lower surface of the hopper can be adjusted. The process begins by adjusting the disc at a considerable distance from the hopper. When the paint has once gone through the apparatus the disc is raised so that the colour is more finely ground. The ground paint is returned to the hopper until it is quite uniform.

This form of mill can be made in different sizes. Fig. 42 shows the construction of a form for grinding by hand. The actual grinding arrangements are exactly the same as those previously described; the difference lies simply in the use of a fly-wheel turned by a handle. The illustrations are due to the kindness of W. Sattler of Schweinfurt, who makes paint mills as a speciality in different sizes and of excellent quality.

In the second form of paint mill the paint is passed

Fig. 42.

between smooth rollers moving in opposite directions with different speeds, which thus exert a grinding action in addition to the crushing effect. The two rollers are provided with cog-wheels with a different number of cogs in order to give the different speeds.

In order that the paint may be sufficiently ground in one operation, roller mills are made with several pairs of rollers, one above the other, the lower rollers being somewhat nearer together than the upper. The paint is fed on to the top pair of rollers, and, after going through these, passes to the next,

THE PREPARATION OF PIGMENTS FOR PAINTING. 433

and finally, after going between all the pairs of rollers, collects in a receiver below in a finished condition. By the use of such mills the paints can be ground sufficiently fine in one operation, if a proper number of rollers is used. When only two or three pairs are used and the paint is to be very finely ground, it must be passed through the whole mill two or three times.

CHAPTER LXX.

THE EXAMINATION OF PIGMENTS.

WHEN an accurate examination of a pigment is required, the only course is to conduct an exact chemical analysis, which can only be done by an expert chemist in a well-equipped laboratory. But often it is necessary in the course of trade to decide rapidly the nature of a pigment or to detect the adulteration of a dear pigment with a cheaper, and for this purpose it is important to have a method of examination which can be conducted without much apparatus, and which requires no great chemical knowledge. It is quite possible to test the majority of pigments in a simple manner. Few reagents are necessary. For an examination of the mineral pigments the following are generally sufficient: hydrochloric, nitric and sulphuric acids, caustic soda solution and ammonium sulphide.

The examination of a pigment containing organic compounds is somewhat more difficult, especially when it is necessary to ascertain the nature of the colouring matter. In this case additional reagents are required—stannous chloride, alum solution, etc.

Since the present work is intended to meet the requirements of the practical man, the behaviour of the different pigments towards ordinary reagents is given in tabular form, the pigments of the same colour being taken together. Since the colouring matters of organic origin, as they occur in the lakes, require a rather more detailed examination, the testing of pigments composed of inorganic materials only is first given, afterwards the properties of the organic colouring matters, so far as is necessary, will be added.

Mineral Pigments.—If the substance to be examined is in the form of a dry powder, it may at once be tested with the reagents mentioned, but if it is a paint or water colour, the oil or gum must be first removed, otherwise it would not be possible to recognise with certainty the action of the reagents.

From water colours, which are ground with gum Arabic or tragacanth solution, it is easy to separate the pigment from the binding medium. The colour is allowed to stand in a tall narrow beaker with a somewhat large quantity of water. After some time the lumps become soft, they are then repeatedly stirred up, the pigment allowed to deposit, and the water, which now contains the binding medium, poured off.

The removal of oil from a paint is somewhat more difficult. A quantity of the paint is placed in a flask with a mixture of equal parts of strong alcohol and ether, or with benzine. The flask is lightly corked, and, after frequent shaking, allowed to stand. The liquids mentioned are good solvents for oils. After a day or two the pigment will generally have deposited at the bottom of the flask. The solution is then poured off, the residue mixed with a small quantity of the solvent and transferred to a filter, where another small quantity of solvent is poured over it. The solvent is allowed to drain off, and the residue is dried. A powder without coherence should be left; this is the pigment free from oil. It can be treated like an originally dry colour.

The pigments are most conveniently examined in test tubes. If these are not at hand the reactions may be carried out on a plate of glass lying on white paper. The powder is placed on the glass and the reagent dropped on to it from a glass rod dipped in the liquid.

Examination with the Blowpipe.—From the behaviour of pigments at high temperatures important conclusions may be drawn as to their nature. For this purpose small porcelain crucibles are used; broken pieces of porcelain or an iron spoon may also be employed. An ordinary spirit lamp is generally sufficient as a source of heat. In some cases a higher temperature is required, which is obtained by means of the blowpipe.

The blowpipe is an invaluable instrument in the examination of mineral pigments. By means of it, almost without reagents, the nature of the pigment can generally be ascertained. The reagents necessary in using the blowpipe are soda, borax, and the solution of a cobalt salt.

The following method should be observed in examining pigments by means of the blowpipe. A quantity of the substance, about equal in volume to two grains of rice, is placed in a small hole cut by a knife in a piece of charcoal, where it is heated by the blowpipe flame. The behaviour of substances is different in the oxidising and reducing flames of the blowpipe. When a flame is blown out to a point by means of the blowpipe it may be seen that the flame consists of two conical portions, one inside the other. The inner is known as the reducing flame, because metallic oxides heated in it produce a bead of metal, or, as the change is chemically expressed, are reduced to metal. The outer cone has the opposite properties; metals melted in it are quickly changed into oxides by the action of the oxygen of the air, which has unrestricted access. In examining a pigment with the blowpipe the reducing flame is first used. The nature of the bead of metal, such as is readily obtained from lead pigments, often allows the composition of the substance to be recognised with certainty. If the behaviour of the bead of metal is not conclusive it is further heated in the oxidising flame. The metal is thus converted into oxide, which deposits on the charcoal, and by its colour and volatility or want of volatility enables the metal contained in the pigment to be determined.

A solution of cobalt nitrate or chloride is used in testing for certain metallic oxides. The substances are moistened with a very dilute solution of one of these substances before heating. After they have been heated they show characteristic colours if certain oxides are present.

Several metallic oxides give a characteristic colour when fused with borax. For this purpose a small loop is made at the end of a thin platinum wire; this is moistened and dipped in powdered anhydrous borax. On heating in the blowpipe flame the borax adhering to the wire melts to a colourless glass. In testing a pigment the transparent bead of borax is then dipped into the powder, and again fused in the blowpipe flame. It is of great importance in this test to fuse but a very small quantity of material in the borax bead; some metallic oxides have such great colouring power that, when too much is used, the bead appears quite black, so that its colour cannot be recognised.

Reactions of the White Pigments.

On Heating on Charcoal :—Lead pigments give a lead bead in the reducing flame, which is converted in the oxidising flame to lead oxide, forming a deposit on the charcoal surrounding the hole.

Antimony white gives a brittle metallic bead in the reducing flame. This burns in the oxidising flame with the evolution of white vapours, and is at the same time covered by small shining crystals.

Bismuth white in the oxidising flame gives a rainbow-coloured incrustation spreading far over the charcoal.

Tin white gives a malleable metallic bead, converted by nitric acid into a white powder.

Zinc white is converted into a green mass when moistened with cobalt solution and heated in the oxidising flame.

The more expensive white pigments are frequently mixed with cheaper substances, *e.g.*, white lead with finely powdered barytes, chalk or gypsum. Such admixture is distinctly to be regarded as adulteration, since the added substances have not the covering power of white lead. In the case of coloured pigments an addition of a white material is not to be regarded as adulteration, since the addition is made in order to impart a paler shade to the colour.

Pigment.	Hydrochloric Acid.	Caustic Soda.	Ammonium Sulphide.	On Heating
Antimony white	Dissolves, solution turbid on adding water.	Dissolves.	Turns reddish yellow.	Turns yellow and melts.
White lead	Dissolves with effervescence, solution gives crystalline lead chloride.	Dissolves.	Turns black.	Becomes permanently yellow.
Lead oxychloride	Dissolves without effervescence.	Dissolves on boiling.	Turns black.	Turns yellow.
Lead sulphate	Insoluble.	Dissolves on boiling.	Turns black.	Unaltered.
Permanent white (barium sulphate)	Insoluble.	Insoluble.	Unaltered.	Unaltered.
Bismuth white	Insoluble.	Insoluble.	Turns black.	Evolves reddish brown fumes, which redden litmus paper.
Zinc white	Dissolves without effervescence.	Dissolves.	Unaltered.	Turns yellow, becomes white again on cooling.
Tin white	Dissolves.	Dissolves.	Turns yellow.	Unaltered.

REACTIONS OF THE YELLOW PIGMENTS.

Pigment.	Hydrochloric Acid.	Caustic Soda.	Ammonium Sulphide.	On Heating.
Chrome yellow and chrome red	Green solution above a white residue, which is soluble on largely diluting.	Becomes orange on boiling and dissolves.	Blackened.	Fuses to a yellow mass.
Cassel yellow . .	Unaltered, becomes white on boiling.	Becomes paler on boiling, liquid yellow.	Blackened.	Melts.
Naples yellow .	On boiling orange, then white.	Reddish yellow.	Turns brownish black.	Melts at a high temperature.
Massicot . . .	Turns white.	Partially soluble on boiling.	Blackened.	Melts with some difficulty.
Lead iodide . .	Turns white.	Dissolves.	Blackened.	Melts.
Barium yellow (chromate)	Yellow solution which gives white precipitate with sulphuric acid.	Unaltered.	Unaltered.	Unaltered.
Cadmium yellow	Dissolves with evolution of sulphuretted hydrogen.	Unaltered.	Unaltered.	Melts with difficulty.
Zinc yellow . .	Yellow solution.	Yellow solution, white residue.	Unaltered.	Melts with difficulty.
Cobalt yellow . .	Red solution.	Colourless solution, greyish blue precipitate.	Unaltered.	Becomes blackish at high temperatures.
Orpiment . . .	Unaltered.	Colourless solution.	Yellow solution.	Volatilises.
Turpeth mineral	Dissolves.	Unaltered.	Blackened.	Turns red.

On Heating on Charcoal:—Chrome yellow, chrome red, Cassel yellow, massicot and lead iodide give lead beads in the reducing flame. Chrome yellow and red when fused with soda give red masses soluble in water. Naples yellow gives a lead bead and white fumes without smell.

Orpiment gives an odour of garlic.

Cobalt yellow, when heated with alumina, is turned blue.

Cadmium yellow produces a brown incrustation on the charcoal.

The examination before the blowpipe serves especially for the recognition of lead, chromium, antimony and arsenic. If a pigment blackens on heating, the presence of an organic colouring matter is indicated, such as Dutch pink, weld lake, etc.

The adulteration of pigments free from lead by colours containing that metal is recognised by the blackening produced by ammonium sulphide.

REACTIONS OF THE RED PIGMENTS.

Pigment.	Hydrochloric Acid.	Caustic Soda.	Ammonium Sulphide.	On Heating.
Chrome red	Green solution, white residue soluble on largely diluting.	Yellow solution and white residue.	Turns greenish black.	Fuses.
Red lead	Chlorine is evolved, white residue.	Almost unchanged.	Turns black.	Turns yellow and finally melts.
Ferric oxide pigments	Slowly dissolve to yellow solution.	Unaltered.	Slowly blackened.	Become dark blackish brown.
Antimony vermilion	Dissolves with evolution of sulphuretted hydrogen.	Dissolves to colourless solution.	Becomes darker, partially soluble.	Melts.
Mercury vermilion	Unaltered.	Turns yellowish.	Unaltered.	Volatilises, sulphur dioxide evolved.
Mercuric iodide	Dissolves to colourless solution.	Dissolves to yellowish solution.	Blackened.	Fuses and then volatilises.
Realgar	Unaltered.	Dissolves to colourless solution.	Dissolves to yellow solution.	Volatilises.

On Heating on Charcoal:—Chrome red and red lead give lead beads in the reducing flame. The former gives a red mass when fused with soda, which dissolves to a yellow solution.

Ferric oxide pigments become darker, but give no incrustation.

Antimony vermilion burns with production of sulphur dioxide and white fumes without smell when heated in the oxidising flame; when fused with soda before the blowpipe, it gives a white brittle bead of metallic antimony.

Vermilion volatilises in the oxidising flame and gives a smell of sulphur dioxide.

Mercuric iodide readily fuses and volatilises.

Realgar volatilises. When heated with soda in the reducing flame, white fumes with an odour of garlic are produced.

REACTIONS OF THE BLUE PIGMENTS.

Pigment.	Hydrochloric Acid.	Caustic Soda.	Ammonium Sulphide.	On Heating.
Prussian, Chinese, Paris, Turnbull's, and Brunswick blue	Dissolve to green solution, then yellow.	Decolourised, brown residue.	Liquid yellowish green.	Blackened.
Mountain blue	Dissolves to yellowish green solution.	Blackened.	Blackened.	Blackened.
Ultramarine	Rapidly decomposed with evolution of sulphuretted hydrogen.	Unchanged.	Unchanged.	Unchanged.
Smalts	Almost unaltered, greenish solution on long boiling.	Unchanged.	Blackened.	Fuses at a high temperature.
Cobalt blue	Unchanged.	Unchanged.	Unchanged.	Infusible and unchanged.

On Heating on Charcoal:—Prussian, Chinese, Paris, Turnbull's and Brunswick blue are turned black, the residue colours the borax bead pale brown in the oxidising flame, and pale green in the reducing flame.

Ultramarine is unaltered at a high temperature.

Smalts, on long heating in the reducing flame with borax, gives a dark blue bead.

Cobalt blue is infusible; it colours the borax bead blue. The bead loses its fine colour on long heating in the reducing flame.

Mountain blue is blackened before the blowpipe. When the residue is moistened with hydrochloric acid and again heated, the flame is coloured bright green. When fused with borax in the oxidising flame, an emerald green bead is formed.

REACTIONS OF THE GREEN PIGMENTS.

Pigment.	Hydrochloric Acid.	Caustic Soda.	Ammonium Sulphide.	On Heating.
Verdigris . . . (all varieties)	Dissolves to green solution, smell of acetic acid.	Unaltered.	Blackened.	Blackened with evolution of peculiar odour.
Bremen green and Brunswick green	Green solution and white residue.	Unaltered.	Blackened.	Blackened.
Emerald green, Scheele's green	Dissolve to greenish solution.	Gradually coloured brownish yellow.	Become brownish black.	Blackened and evolve garlic-like odour.
Copper borate .	Dissolves to greenish solution.	Black residue.	Becomes brownish black.	Fuses.
Rinmann's green	Dissolves to rose red solution.	Unaltered.	Blackened.	Unaltered.
Chromium oxide	Almost unaltered.	Unaltered.	Becomes dark dirty green.	Unaltered.
Chrome green lake	Becomes deeper in colour.	Unaltered.	Becomes dark dirty green.	Unaltered.
Manganese green	Dissolves to green solution.	Dissolves to green solution.	Is discoloured.	Unaltered.
Green ultramarine	Is decolourised with evolution of sulphuretted hydrogen.	Unaltered.	Unaltered.	Unaltered.

On Heating on Charcoal:—Verdigris, Bremen and Brunswick greens give black residues on charcoal, which produce a bluish-green bead when fused with borax in the oxidising flame.

Emerald green and Scheele's green behave in a similar manner, but on heating evolve an odour of garlic.

Rinmann's green gives a blue borax bead.

Manganese green is discoloured in the reducing flame.

REACTIONS OF THE BROWN PIGMENTS.

Pigment.	Hydrochloric Acid.	Caustic Soda.	Ammonium Sulphide.	On Heating.
Lead brown	White residue, chlorine evolved.	Dissolves.	Blackened.	Turns yellow and fuses.
Manganese brown	Dissolves to yellow solution.	Unaltered.	Becomes flesh coloured.	Unaltered.
Pyrolusite brown	Dissolves to yellow solution, chlorine evolved.	Unaltered.	Becomes flesh coloured.	Unaltered.
Prussian brown	Dissolves to yellow solution.	Unaltered.	Blackened.	Turns to reddish brown.
Iron brown	Dissolves to yellow solution.	Unaltered.	Blackened.	Unaltered.
Chrome brown	Dissolves to greenish yellow solution.	Yellow solution, black residue.	Blackened.	Blackened.
Cobalt brown	Dissolves to reddish yellow solution.	Unaltered or blackened.	Blackened.	Unaltered.
Hatchett brown	Unaltered.	Becomes greenish blue.	Blackened.	Blackened.
Humins, Bistre	Unaltered, yellow liquid.	Give yellow liquid.	Unaltered.	Burn.

On Heating on Charcoal:—Lead brown gives a lead bead in the reducing flame.

Manganese brown and pyrolusite brown, when fused with saltpetre on platinum foil at a high temperature, give a bluish green mass.

Prussian brown and iron brown give a pale green borax bead in the reducing flame, which turns yellowish brown in the oxidising flame.

Chrome brown heated, moistened with hydrochloric acid, and again heated, colours the flame green. It also gives a green borax bead.

Cobalt brown produces a blue borax bead.

Humin substances burn when heated on charcoal.

REACTIONS OF THE BLACK PIGMENTS.

Almost all black pigments consist of carbon, upon which reagents have no action. They should be at once heated on charcoal. If they burn away completely in the oxidising flame they consist of lamp-black or carbon obtained by some process of incomplete combustion; if a white, infusible residue is left, the pigment is bone (ivory) black; if the residue is black the substance under examination must be "neutral tint," chrome black, or chrome-copper black. The two former give a pale green borax bead, whilst chrome-copper black gives a deep green bead, and when heated, moistened with hydrochloric acid, and again heated, it colours the flame green.

CHAPTER LXXII.

THE TESTING OF DYE-WOODS.

In the manufacture of pigments from dye-woods or other organic substances, the value of the raw material is in proportion to the amount of colouring matter it contains, other things being equal. It is specially desirable to estimate accurately the colouring matter in expensive materials such as indigo and cochineal.

There are a number of methods which permit an accurate estimation of the indigo blue in indigo. One good process is founded upon the decomposition of indigo blue by chlorine, when the colour of the solution changes from blue to yellow. Since a definite amount of chlorine is required to decompose indigo blue, from the quantity of chlorine required by a sample of indigo, its content in indigo blue can be ascertained.

Whilst the percentage of indigo blue contained in indigo can be found with tolerable accuracy, though by a rather elaborate process requiring special apparatus, there is no convenient method for examining the other organic colour materials by which their content in active constituents can be readily found. In practice a process is particularly valuable which requires little time and no complicated apparatus. Colouring materials can be [rapidly tested by a physical process which requires little time and an inexpensive apparatus. Under similar conditions the extract of a dye-wood is deeper in colour in proportion to the colouring matter it contains. If therefore the intensity of the colour of the extract can be accurately measured, there is no difficulty in drawing a certain conclusion as to the amount of colouring matter in the raw material.

The Colorimeter is the apparatus adapted for the purpose

Fig. 43.

in question. There are many forms of colorimeter. The instrument devised by Dubosq is distinguished by simplicity and accuracy of results before other apparatus of similar construction. Dubosq's colorimeter consists of the following parts (Fig. 43): two glass cylinders, C and C_1, the bottoms of which must be perfectly plane both inside and outside (the accuracy of the results depends upon this), stand upon a sheet of plate glass. Two glass cylinders of smaller diameter, T and T_1, are suspended in C and C_1. The bottoms of these cylinders must also be quite plane. It would be very expensive to make glass cylinders of this kind in one piece. The same result is obtained by providing each cylinder with a metal ring upon which screws another ring in which is cemented a circular piece of plate glass.

Light should only reach the eye of the observer in a direction parallel with the axis of the cylinders. C and C_1 are therefore blackened on the outside. The inner cylinders, T and T_1, are fastened to racks moving vertically. The distance through which the cylinder is moved is measured by a scale on one of the racks. Above the cylinders T and T_1 are Fresnel's prisms. Below C and C_1 is a mirror, S, which can be adjusted to throw light vertically upwards. Beams of light pass through the plate glass and the bottoms of the cylinders C and T, C_1 and T_1, unrefracted, they are then deviated by the Fresnel's prisms so that the observer looking down through the telescope, F, has a circular field of view, one half of which is illuminated by the light passing through the cylinder C, and the other by the beam passing through C_1. The intensity of the light which has passed through the two cylinders can thus be accurately compared.

In order to use this apparatus to compare the intensity of colour of two liquids, the following process is performed: A liquid is made, the colour intensity of which is taken as 100. The colour intensity of the liquid under examination will then be represented by a number indicating the relation between its intensity and that of the standard liquid. A solution of caramel in water is generally used as the standard, since this substance has very great intensity of colour. The preparation of absolutely pure caramel is difficult, and it is therefore advisable, in order always to have the standard solution of the same intensity, to prepare a large quantity of caramel solution at once, to add carbolic acid to prevent it from decomposing, and to keep it in well-closed bottles. When the standard solution is almost used up the colorimeter is employed to prepare a fresh quantity of equal intensity.

To render possible an exact comparison of two substances they must be tested under the same conditions, that is, the solutions of the colouring matters must be made in exactly the same manner. Finely-powdered materials dissolve more readily than coarse powders. In making the solutions which are to be examined for intensity of colour, the raw materials must be brought into a condition of fine division by the same instrument, for example, a rasp. The colour solution is made by boiling 100 grammes of the dye-wood with exactly a litre of distilled water for precisely thirty minutes. The liquid is then filtered into a 1-litre flask. The dye-wood absorbs water, and some is lost by evaporation, so that considerably less than 1 litre of liquid is collected in the flask. Distilled water is added to make up the volume to 1 litre.

If two samples of logwood are treated in this manner, solutions are obtained which contain the colouring matter in the same proportions in which it exists in the two samples. The intensity of colour of the solutions is then estimated in the following manner: The cylinder C is filled with the standard caramel solution up to a mark on the outside of the cylinder. The cylinder C_1 is filled with the decoction to the same height. The distance between the bottoms of the cylinders, R and T, R_1 and T_1, must be made smaller in proportion to the intensity of colour of the liquid between them. If now the amount of light which penetrates a layer of caramel solution of a certain thickness be taken as unity, the depth of a layer of the decoction must be greater, the smaller the quantity of colouring matter it contains, in order that the two halves of the field of view may be equally illuminated. The cylinder T_1 must be raised to a greater height the smaller the quantity of colouring matter in the liquid, in order that the two halves of the field may be illuminated to the same extent. If the colouring power of the caramel solution is taken as 100, the colouring power of the decoction can be readily calculated from the height to which the cylinder, T_1, is raised. The heights of the layers of liquid between the bottoms of the cylinders C and T, C_1 and T_1, are inversely as the quantities of colouring matter contained in the respective cylinders.

When caramel solution is used as the standard the intensity of the light in the two halves of the field can be judged, but not the intensity of colour. In order to estimate the latter a solution of that colouring matter must be used as a standard which is the principal constituent of the decoction under examination. Thus, in a careful examination of logwood a solution of hæmatoxylin would be used, and in the examination of red wood a solution of pure brasilin, as the standard. In this case a saturated solution of the colouring matter would be taken as the standard. If the intensity of its colour were represented by 100, the colour intensity of the wood under examination would always be less than 100, and would, with tolerable accuracy, represent the percentage of colouring matter in the wood. The result would not be quite accurate, because the dye-wood contains other substances which dissolve in water on boiling and affect the colour of the decoction, but the results are of such accuracy that for practical purposes no material mistake will be made by taking them as percentages of colouring matter.

Although at first sight the estimation of the colouring matter in a dye-wood by means of the colorimeter appears somewhat complicated, yet it yields the most accurate results with the smallest expenditure of labour and time. The value of a colouring material may also be estimated by preparing the pure colouring matter from a weighed quantity. This process is lengthy, demands considerable practice, and only gives good results when it is carried out with the most painful accuracy. The colouring matters in question are precipitated by lead salts. If the dye-wood extract contained colouring matter alone its amount could be found by observing the volume of a lead solution of known strength required to precipitate the colouring matter completely. The decoctions, however, contain other substances which form lead compounds, and are precipitated together with the colouring matter, so that if the precipitate were regarded as the pure lead compound of the colouring matter a very inaccurate result would be obtained. In order to obtain results with some pretensions to accuracy the lead compound of the colouring matter must be purified. The impure precipitate is washed and suspended in water, through which sulphuretted hydrogen is passed until all the lead is precipitated as lead sulphide, which is filtered off, excess of sulphuretted hydrogen expelled by boiling, and the solution again precipitated by a lead salt. This precipitate may be regarded without considerable error as the lead compound of the colouring matter. The weight of colouring matter contained in the quantity of wood used can be calculated from the weight of the dry precipitate. This method is somewhat complicated and tedious; the results are inferior in accuracy to those obtained by means of the colorimeter, which instrument furnishes the most suitable method for testing dye-woods for practical purposes.

A thorough knowledge of chemistry is indispensable to the colour manufacturer who wishes to carry on his business on any extensive scale. It enables him to match any sample of colour submitted to him and to test his raw materials with ease. We have indeed given in a section of this book simple methods by which the majority of commercial pigments can be tested with tolerable accuracy by means of a few reagents, and for ordinary purposes these methods are sufficient. But when an accurate examination of a pigment is required, it must be conducted by the ordinary processes of analytical chemistry. The colour manufacturer has not only to carry out these occasional examinations, but has frequently to test certain raw materials which he uses in large quantities. Soda may be taken as an example. An estimation of the percentage of sodium carbonate in this substance is an exceedingly simple matter to the chemist, but can hardly be carried out without a knowledge of chemistry.

A manufacturer without chemical knowledge, who is carrying on an industry which, like the manufacture of colours, rests entirely upon a chemical foundation, will constantly be compelled to seek advice from a scientific chemist. Many pigments can be made according to a settled formula, but the results of working strictly according to the formula, without a knowledge of the reasons for the operations, can only be satisfactory whilst no irregularity occurs. The least irregularity places those who work blindly in a completely helpless position, for they do not know what is wrong and cannot remove the hindrance.

In the manufacture of pigments certain by-products are produced. These can be utilised by a manufacturer possessed of chemical knowledge, whilst they are simply thrown away by many, thus making the manufacture of the particular pigment far more expensive than when the by-product is also made valuable. Strictly speaking, there are no worthless by-products in making pigments every liquid obtained in precipitating a colour might be further utilised. Salt solutions can only be regarded as worthless by-products when the cost of separating the salt from the solution would be greater than the value of the product. Thus we cannot conclude this section of the work without again insisting that the study of chemistry is indispensable to the colour maker, since his industry is chemical from beginning to end. The colour maker who works simply by recipes will never raise himself above the position of an ordinary workman, who does what he is told without thinking of what he is doing. The smallest mistake in carrying out the process generally results in the complete failure of the whole operation and thus causes the manufacturer material loss.

CHAPTER LXXIII.

THE DESIGN OF A COLOUR WORKS.

In the establishment of a colour works several conditions are necessary. The most important is the supply of water in sufficient quantity and purity. It has already been stated that many pigments cannot be made with water containing much organic matter or salts, since the dissolved substances affect the shade. This action not only takes place in the formation of the colour, but is also unpleasantly manifest when it is washed. The delicate lakes are discoloured by organic matter in the water, and are so changed by any considerable quantity of lime that the alteration in shade is clearly perceptible after continued treatment. If the water contains but a very small quantity of iron, the preparation of some pigments is made quite impossible, since—and this is especially the case with the lakes—the ferric oxide is precipitated together with the pigment, and in consequence of its characteristic colour imparts to it an ugly shade. Thus in choosing a site for a colour works the available water supply must be carefully examined. If it is not sufficiently pure or in sufficient quantity, the position must be regarded as unsuitable.

No colour maker, although working on the largest scale, is in a position to make all the materials he requires. With the continual development of the chemical industries, the number of these substances which can be made economically in the colour works continually diminishes. It is far more advantageous to obtain them from works in which they are made on the large scale. Many of these substances are required in large quantity, so that a site should be chosen in direct communication with the railway, so that the cost of carriage is diminished, and also the cost of distributing the manufactured materials. This is especially necessary for materials which have a low value and consequently can bear no high cost of carriage.

In regard to the space required by a colour works, no actual dimensions can be given, since these vary with the extent of the business, and with the pigments produced. In the price lists of large colour works all the commercial pigments are generally quoted, but they are rarely if ever actually made in one works, but are obtained at a lower price from other establishments, which make a speciality of certain pigments, and by working on a large scale can produce them at such a cost that the smaller manufacturer is not in a position to compete with them. Thus there are works in which only white lead or ultramarine is made.

If a colour manufacturer is in the fortunate position of placing his works on a river, he has not only an unrestricted water supply, but may also be able to use water power, the cheapest of all powers. When water power is available, it will be used to raise the large volumes of water required and to move the machinery for grinding the raw materials, rasping dye-woods, etc. In a colour works of any size a boiler is required; if water power is not used, it must be of sufficient power to give steam for driving the engine, for boiling liquids and for heating the drying stoves. The boiler is of great advantage in providing steam for dissolving salts, extracting dye-woods and boiling liquids. When liquids are boiled by steam there is considerable economy in that the majority of the boiling tubs can be of wood, which is provided with a protective coating for liquids which attack wood. Thus there is economy in dispensing with large metal pans and with the fire-places in which they would be built, and also in the course of time there is considerable saving in fuel.

For a well-equipped colour works a drying stove, in which the pigments can be thoroughly dried, is important. It is most convenient to heat drying stoves with steam. The temperature can be easily regulated by increasing or diminishing the supply of steam. For drying pigments which would not be injured by considerable increase of temperature, the drying stove may be heated by a fire.

In a colour works in which many pigments are made sulphuretted hydrogen is frequently required. Since all lead pigments are blackened by this gas, the greatest care is required in using it. Also the poisonous nature of sulphuretted hydrogen is generally under-estimated. In working with sulphuretted hydrogen, the apparatus depicted in Fig. 3 should be used. It should be placed in a position, such as a closed yard, in which escaping gas will be harmless. If this cannot be done the precipitation should be accomplished in closed vessels, from which a pipe should carry the gas to a fire, where it will be burnt to sulphur dioxide and water.

In most cases the manufacturer of colours makes but a certain number of pigments and rarely or never all which are mentioned in his price lists. The dimensions of the establishment will be in accordance. In commencing a new colour works it is advisable from purely financial reasons to produce at first a limited number of colours, but these in perfection. By many experiments and diligent study of chemistry a colour manufacturer may hope to compete successfully in so difficult a branch of chemical technology as the manufacture of colours.

CHAPTER LXXIV.

COMMERCIAL NAMES OF PIGMENTS.

IN commerce the pigments are found under the most different names, the most common of which have been given together with the description of the pigment. No regularity can be found in the names chosen for the different pigments; quite arbitrary designations have been taken. Pigments are most commonly named after places—for example, Prussian blue, Paris blue, Bremen green; also after the discoverer—Turnbull's blue, Hatchett brown. Whilst these names give the place of production or the name of the discoverer, and thus have some foundation, there are many others for which no reason can be assigned, e.g., King's yellow. Certain names are based upon the chemical composition of the pigment. These should be used by preference, but now that the expressions white lead, chrome yellow and Chinese blue have become common no one would think of speaking of basic lead carbonate, lead chromate or ferric ferrocyanide. The confusion in the nomenclature of colours is increased by placing pigments which possess English names upon the market under French, German or Latin names, which are often sadly mutilated. This is more the case in Germany than in England.

It may easily happen that a reader of a work on colour making might search in vain for a pigment whose name he had somewhere heard, whilst the book contained a description of the colour and its properties, but under another name. To remove this difficulty it has been thought necessary to collect the names of the different pigments, which are contained in the following table. The French and German names are also given. The most usual names are printed in italics.

White Pigments.

Basic Lead Carbonate.—*White lead, flake white.*
    Céruse, blanc de plomb, blanc d'argent, blanc de neige, fleur de neige, blanc de Venise.
    Bleiweiss, Schieferweiss, Schneeweiss, Silberweiss, Kremserweiss, Kremnitzerweiss, Berlinerweiss, Venetianerweiss, Hollanderweiss, Hamburgerweiss.

Lead Oxychloride.—*Pattison's white lead.*
    Blanc de Pattison, blanc d'Angleterre.
    Bleiweiss, Pattisonweiss, englisches Patentweiss.

Lead Sulphate.—Lead bottoms.
    Céruse de Mulhouse.
    Bleiweiss, Vitriolweiss.

Barium Sulphate.—*Enamel white*, permanent white, blanc fixe, baryta white.
    Blanc fixe, blanc permanent.
    Permanentweiss, Barytweiss, Schneeweiss, Mineralweiss, Neuweiss.

Zinc Oxide.—*Zinc white*, permanent white, flowers of zinc.
    Blanc de zinc, oxyde de zinc, fleur de zinc.
    Zinkweiss, Zinkblumen, weisses Nichts, Ewigweiss.

Basic Bismuth Nitrate.—*Pearl white*, bismuth white, Spanish white.
    Blanc d'Espagne, blanc de bismuth, blanc de fard.
    Wismuthweiss, Spanischweiss, Perlweiss, Schminkweiss.

YELLOW PIGMENTS.

Lead Chromate.—*Chrome, Chrome yellow.*
   Jaune de chrome, jaune d'or.
   Chromgelb, Königsgelb, Citronengelb, Neugelb, Parisergelb, Leipzigergelb, Kölnergelb, Zwickauergelb, amerikanisches Gelb.

Lead Oxide.—*Litharge, massicot.*
   Bleiglätte, Glätte, Massicot.

Lead Oxychloride.—Patent yellow.
   Jaune minéral, jaune breveté, jaune de Montpellier.
   Casselergelb, Veronesergelb, Mineralgelb, Patentgelb, Englischgelb, Parisergelb, Montpelliergelb.

Lead Antimoniate.—*Naples yellow.*
   Jaune de Naples, jaune d'antimoine.
   Neapelgelb, Antimongelb.
   Giallolino, Giallo di Napoli.

Barium Chromate.—*Lemon yellow*, baryta yellow, yellow ultramarine, permanent yellow.
   Barytgelb, gelbes Ultramarin, Chromgelb.

Zinc Chromate.—*Zinc yellow*, zinc chrome.
   Jaune de zinc, jaune permanent, jaune bouton d'or, jaune de chrome inaltérable, jaune d'outremer.
   Zinkgelb, Chromgelb.

Cadmium Sulphide.—*Cadmium yellow*, aurora yellow.
   Jaune brillant.
   Cadmiumgelb.

Basic Mercuric Sulphate.—*Turpeth mineral*, mercury yellow.
   Jaune de mercure.
   Mercurgelb, Königsgelb, mineralischer Turpeth.

Cobalt Potassium Nitrite.—*Aureolin*, cobalt yellow.
   Jaune indien.
   Kobaltgelb.

Arsenic Disulphide and Trisulphide.—*Realgar*, orpiment, *King's yellow*.

Orpiment, jaune royal.

Realgar, Auripigment, Rauschgelb, Rauschroth, Königsgelb, Chinagelb, Persischgelb, Spanischgelb.

Stannic Sulphide.—Mosaic gold.

Musivgold.

YELLOW LAKES.

Dutch pink, French yellow, *stil de grain*, jaune d'Avignon, jaune français, *Schüttgelb*.

Yellow lake, jaune de gauche, *Waulack*.

Indian yellow, jaune indien, *Purre, Indischgelb*.

Gamboge, gomme-goutti, *Gummigutt, Gummigutti*.

RED PIGMENTS.

Basic Lead Chromate.—*Chrome red, orange chrome, Persian red*, Derby red, Chinese red.

Jaune d'or, *jaune orange*, pâte orange.

Chromroth, Chromorange.

Mercuric Sulphide.—*Vermilion*, cinnabar.

Vermillon, cinabre.

Zinnober, Vermillon, Chinesischroth, Patentroth.

Mercuric Iodide.—*Scarlet*.

Ecarlate.

Jodinroth, Scharlachroth.

Ferric Oxide.—*Rouge, colcothar, Indian red, Venetian red*, crocus, caput mortuum, Mars red.

Rouge des Indes, rouge de Mars, rouge d'Angleterre.

Englischroth, Engelroth, Berlinerroth, Königsroth, Kaiserroth, Neapelroth, *Indischroth*, Persischroth, Eisensafran, Todtenkopfroth, *Marsroth*.

Cobalt Phosphate or Arsenate.
   Chaux métallique.
   Kobaltrosa.

RED LAKES.

Cochineal Lakes.—*Carmine, crimson lake.*
   Laque carminée.
   Carmin, Cochenilleroth, Münchenerlack, Wienerlack, Florentinerlack, Pariserlack.

Lac Dye.
   Lack-lack, Lack-dye.

Madder Lakes.—*Rose madder*, purple madder, madder carmine.
   Laque de garance.
   Krapplack, Wienerlack, *Krappcarmin*, Garancincarmin.

Red Wood Lakes.—Rose pink, Florentine lake.
   Laque en boules, laque de Vienne, laque de Venise, laque de Florence.
   Kugellack, Münchenerlack, *Wienerlack, Berlinerlack, Florentinerlack*, Venezianerlack, Neulack.

Carthamine.—Safflower, Spanish red.
   Rouge de carthame, rouge de Chine, rouge d'or, rouge en écailles, rouge végétal, rouge de Portugal.
   Safflorcarmine, Safflorroth, Tassenroth, *Tellerroth*, Vegetalroth.

BLUE PIGMENTS.

Ferric Ferrocyanide.—*Prussian blue, Chinese blue*, Paris blue, Berlin blue, *Brunswick blue.*
   Bleu de Prusse, bleu de Paris, *bleu de Berlin*, bleu d'Anvers.
   Pariserblau, Berlinerblau, Preussischblau, Sächsischblau, Neublau, Oelblau, Wasserblau, Mineralblau, Erlangerblau, Zwickauerblau, Waschblau, Louisenblau, Raymondblau.

Ferrous Ferricyanide.—Turnbull's blue.
  Bleu de Turnbull.
  Turnbull's Blau.
  (The names given under ferric ferrocyanide are also used.)

Ultramarine.—Lime blue, Royal blue.
  Outremer, bleu d'azur.
  Ultramarin, Azurblau, Lazurblau, Lapis lazuli-Blau.

Copper Hydroxide or Carbonate (singly or together).—*Blue verditer*, lime blue, mountain blue.
  Bleu de montagne, bleu de chaux, bleu de cuivre, cendres bleues, bleu de Payen.
  Bergblau, Mineralblau, Oelblau, Kalkblau, Neubergblau, Kupferblau, Bremerblau, Steinblau, Hamburgerblau, Neuwiederblau, Casselerblau.

Cobalt Oxide Alumina Compound.—*Cobalt blue*, Thénard's blue.
  Bleu de Thénard, bleu de cobalt.
  Kobaltblau, Thénard'sches Blau, Kobalt-ultramarin, Königsblau, Leydenerblau, Leithnerblau, Wienerblau, Cæruleum.

Cobalt Potassium Silicate.—*Smalts*.
  Bleu de smalt, bleu d'azur, *bleu de Saxe*.
  Smalte, Schmalte, Blaufarbenglas, Sächsischblau, *Streublau*, Königsblau, Kaiserblau, Azurblau, *Eschel*.

BLUE ORGANIC PIGMENTS.

Indigo Sulphonates.—*Indigo carmine*, soluble indigo.
  Bleu de Saxe.
  Indigocarmin, Carminblau, präcipitirter Indigo, *Carminblau*.

Indigo Lake.—Bleu d'Angleterre, bleu de Hollande.
  Neublau, Waschblau, Holländerblau, Englischblau, Tafelindigo.

Green Pigments.

Copper Carbonate.—*Mountain green*, Hungarian green, lime green.

Vert de montagne, vert minéral, vert de cuivre, vert de Hongrie.

Berggrün, Kupfergrün, ungarisches Grün, Tirolergrün, Malachitgrün, Mineralgrün, Schiefergrün, Glanzgrün, Staubgrün, Wiesengrün, Apollogrün, Wassergrün, Oelgrün, Alexandergrün.

Copper Arsenite.—*Scheele's green, verditer*, lime green, *mineral green*.

Vert de Suède, vert de Scheele.

Scheel'sches Grün, *Schwedisches Grün*, Mineralgrün, Braunschweigergrün, Neuwiedergrün, Erdgrün, Aschengrün.

Copper Aceto-Arsenite.—*Emerald green*.

Vert de Vienne, vert de Mitis, vert breveté.

Schweinfurter Grün, Mitisgrün, Wiesengrün, Englischgrün, Patentgrün, Hörmann's Grün, Papageigrün, Kaisergrün, Königsgrün, Wienergrün, Kirchbergergrün, Leipzigergrün, Zwickauergrün, Baslergrün, Parisergrün, Neuwiedergrün, Würzburgergrün, Originalgrün, Jasnügergrün.

Copper Stannate.

Gentele's Grün, Zinngrün.

Copper Oxychloride.—

Kuhlmann's Grün, Elsner's Grün, giftfreies Grün.

Copper Borate.

Borgrün, Kupfergrün, giftfreies Kupfergrün.

Copper Acetate.—*Verdigris*, distilled or crystallised verdigris.

Vert de gris, vert de gris naturel, vert de gris distillé, vert de gris en grappes.

Grünspan, destillirter, französischer, deutscher, präcipitirter or krystallisirter Grünspan.

Chromium Oxide and Hydroxide.—*Guignet's green*, chrome green, lime green, *viridian*.

Vert de chrome, vert Pannetier, vert Guignet, vert de soie, vert émeraude, vert naturel, vert virginal.

Chromgrün, grüner Zinnober, Laubgrün, Smaragdgrün, Deckgrün, Myrthengrün, Permanentgrün, Amerikanergrün, Neapelgrün, Gothaergrün, Guignet's Grün, Mittler's Grün, Pannetier's Grün, Chromgrün in Lack, Türkisgrün, Seidengrün, Naturgrün.

[The green produced by mixing chrome yellow with Prussian blue is known in England as chrome green, Brunswick green, oil green, etc., and on the continent by the names given above for chromium oxide.]

Chromium Phosphate.—Arnaudan's green.

Vert Arnaudan, vert de Plessy.

Arnaudan's, Plessy's, Schnitzer's Grün.

Cobalt Oxide-Zinc Oxide.—*Cobalt green*, Rinmann's green.

Vert de Rinmann, vert de cobalt.

Kobaltgrün, Rinmann's Grün, Zinkgrün, permanenter grüner Zinnober.

Barium Manganate.—Rosenstiehl's green.

Vert tiges de roses.

Mangangrün, *Rosenstiehl's Grün*, Böttger's Grün.

Violet Pigments.

Chromic Chloride.

Chrombronze, Permanentbronze, Tapetenbronze.

Manganese Phosphate.—Nuremberg violet.

Manganviolett, Nürnberger Violett.

Brown Pigments.

Lead Peroxide.

Bleibraun, Flohbraun.

Manganic Oxide and Peroxide.—Manganese brown.

Brun de Mangane, bistre minéral.

Manganbraun, Bisterbraun, *Mineralbister*, Kastanienbraun, *Braunsteinbraun*.

Copper Potassium Ferrocyanide.—Hatchett brown.

Brun de Prusse.

Hatchett's Braun, Kupferbraun, Chemischbraun, Breslauerbraun.

Ferric Oxide.—Brown ochre, Mars brown.

Eisenbraun, Van Dyck Braun, Ockerbraun, Sienabraun.

Prussian brown is Prussian blue decomposed by heat (see page 280).

Chrome brown and cobalt brown (see pages 280, 281).

BLACK PIGMENTS.

Carbon.—*Ivory black*, bone black, Frankfort black, vine black, *vejetable black, drop black, carbon black, lamp black*.

Noir d'ivoire, noir de Frankfort, noir de Cologne, noir d'Allemagne.

Kienruss, Flatterruss, Russschwarz, Rebenschwarz, Hefeschwarz, Beinschwarz, Spodium, Elfenbeinschwarz, Frankfurterschwarz, Pariserschwarz, Wienerschwarz, Lampenschwarz, Oelschwarz, Spanischschwarz, Drusenschwarz.

Carbon mixed with Other Substances.

Neutral tint, composition black, neutral black.

Teint neutre, noir de composition.

Compositionsschwarz, Neutraltinte, Naturaltinte.

Indian ink.

Encre de Chine.

Tusche.

APPENDIX.

THE CONVERSION OF METRIC INTO ENGLISH WEIGHTS AND MEASURES.

The conversion of metric into English weights and measures can be readily accomplished by means of the following relationships :—

MEASURES OF LENGTH.

1 metre = 100 centimetres = 39·37 inches.
1 foot = 0·3381 metre.

MEASURES OF CAPACITY.

1 litre = 1,000 cubic centimetres = ·2209 gall.
1 gall. = 4·5436 litres.

MEASURES OF WEIGHT.

1 kilogramme = 1,000 grammes = 2·205 lb.
1 gramme   = 15·443 grains.
1 cwt.   = 50·802 kilogrammes.

CENTIGRADE AND FAHRENHEIT THERMOMETER SCALES.

To convert temperatures expressed in Centigrade degrees to Fahrenheit degrees multiply by 9, divide by 5, and add 32.

To convert temperatures expressed in Fahrenheit degrees to Centigrade degrees subtract 32, multiply by 5, and divide by 9.

INDEX.

ACID, acetic, 30.
— carbonic, 30.
— hydrochloric, 25.
— nitric, 28.
— oxalic, 31.
— sulphuric, 27.
— tartaric, 31.
Acids, 25.
Alizarin, 372.
Alkalis, 32.
Alkanet, 368.
Alum, ammonia, 48.
— Roman, 46.
— soda, 47.
Alumina, 49.
— gold purple, 193.
Aluminium compounds, 42.
— sulphate, 43.
Alums, 44.
Ammonia, 23.
— cochineal, 362.
Ammonium chloride, 25.
— sulphide, 25.
Annaline, 131.
Antimony blue, 179.
— compounds, 59.
— oxychloride, 129.
— trioxide, 129.
— vermilion, 178.
— yellow, 151.
Antwerp blue, 203.
Apparatus, washing, 120.
Appendix, 469.
Aqua regia, 29.
Archil, 381.
Arsenic compounds, 59.
Asphaltum, 114.
Aureolin, 156.

BARIUM carbonate, 42.
— chloride, 41.

Barium compounds, 41.
— green, Böttger's, 271.
— sulphate, 116.
— yellow, 152.
Barytes, 116.
Bismuth compounds, 59.
— white, 130.
Bistre, 284.
Black, bone, 290.
— charcoal, 286.
— chrome, 318.
— chrome-copper, 318.
— enamel, 325.
— ivory, 290.
— lamp, 307.
— neutral tint, 318,
— pigments, 285, 444.
— pine, 312.
— soot, 294, 313.
— vine, 288.
Blowpipe, 436.
Blue, antimony, 179.
— Antwerp, 203.
— Bremen, 226.
— Brunswick, 200.
— Chinese, 196, 201.
— chrome, 267.
— cobalt, 230.
— copper pigments, 226.
— Egyptian, 250.
— enamels, 323.
— lakes, 390, 397.
— lime, 228.
— manganese, 272.
— mineral, 200.
— mineral pigments, 194.
— molybdenum, 239.
— Neuberg, 227.
— oil, 228.
— Payen's, 228.
— pigments, examination of, 441.

Blue, Prussian, 199.
— Soluble Prussian, 200.
— Tessié du Motay's, 293.
— tungsten, 238.
— Turnbull's, 203.
— ultramarine, 204.
— — pale, 223.
— verdigris, 252.
Bone black, 290.
Böttger's barium green, 271.
Brasileïn, 384.
Brasilin, 384.
Brazil wood, 384.
Bremen blue, 226.
— green, 226.
Brocade pigments, 341.
Bronze, electrolytic copper, 337.
— pigments, 329.
— pigments, tungsten, 338.
Brown, chrome, 280.
— cobalt, 281.
— copper, 280.
— decomposition products, 283.
— Hachett's, 280.
— iron, 280.
— lead, 279.
— manganese, 279.
— mineral pigments, 279.
— organic pigments, 414.
— pigments, examination of, 443.
— Prussian, 280.
— pyrolusite, 279.
Brunswick green, 243.

CADMIUM compounds, 53.
— yellow, 153.
Cœruleum, 231.
Calcium carbonate, 40.
— compounds, 39.
— hydroxide, 39.
— oxide, 39.
— phosphate, 40.
— sulphate, 40.
Carajuru, 379.
Carbon, 30.
Carmine, 354, 357.
— indigo, 394.
— madder, 376.
— safflower, 367.
Carthamine red, 366.
Casselmann's green, 249.
Cassel yellow, 148.
Cassius, purple of, 190.

Caustic potash, 33.
Caustic soda, 37.
Charcoal blacks, 286.
Charvin's green, 412.
Chica red, 379.
Chinese blue, 196, 201.
— green, 411.
— vermilion, 170.
Chlorine, 20.
Chlorophyll, 409.
Chromaventurine, 266.
Chrome alum, 35.
— black, 318.
— blue, 267.
— brown, 280.
— copper black, 318.
— green, 260, 274.
— — Elsner's, 274.
— — lake, 264.
— red, 186.
— yellow, 134.
— — cadmium, 153.
— — calcium, 151.
— — pale, 139.
— — zinc, 152.
Chromic chloride, 276.
Chromium compounds, 58.
— oxide, 260.
— stannate, 189.
Chrysean, 162.
Cobalt arsenate, 189.
— blue, 230.
— brown, 280.
— compounds, 56.
— green, 268.
— magnesia red, 188.
— red, 188.
— ultramarine, 230.
— zinc phosphate, 232.
Cochineal, 354.
Colorimeter, 450.
Colours, simple and mixed, 71.
Colour works, design of, 457.
Commercial names of pigments, 460.
Confectionery colours, 427.
Copper acetate, 65.
— arsenite, 241.
— borate, 250.
— brown, 280.
— carbonate, 240.
— compounds, 65.
— hydroxide, 229.

Copper nitrate, 65.
— oxychloride, 243.
— silicate, 250.
— stannate, 248.
— sulphate, 65.
— violet, 278.
Covering power, 10.
Crayons, 423.
Cream of tartar, 34.
Cudbear, 382.
Curucuru, 379.

DESIGN of a colour works, 457.
Dextrine, 420.
Drying oils, 429.
— stove, 459.
Dutch pink, 348.
Dye-woods, 449.

EGYPTIAN blue, 250.
Elsner's chrome green, 274.
— green, 249.
Emerald green, 244, 264.
Enamel colours, 319.
Enamel white, 116.
Enamels, black, 325.
— blue, 323.
— green, 324.
— red, 322.
— violet, 324.
— white, 320.
— yellow, 322.
Examination of pigments, 434.
Extracts, 398, 402.

FERNAMBUCO wood, 384, 386.
Ferric oxide pigments, 180.
Ferrous chloride, 54.
— sulphate, 54.
Filter press, 124, 125.
Florentine lake, 361.
French purple, 38.
Fustic lake, 351.

GAMBOGE lake, 349.
— prepared, 350.
Garanceux, 372.
Garancin, 371.
Gardinia grandiflora, colouring matter of, 353.
Glucose, 422.
Gold, compounds of, 69.

Green, Arnaudan's, 265.
— Bremen, 226.
— Brunswick, 243.
Green, Casselmann's, 249.
— Charvin's, 412.
— Chinese, 411.
— chrome, 260, 274.
— cobalt, 268.
— Elsner's, 249.
— emerald, 244, 264.
— enamel, 324.
— Guignet's, 264.
— Kuhlmann's, 249.
— lakes, 409.
— leaf, 265.
— lime, 250.
— manganese, 270.
— mineral pigments, 240.
— natural, 275.
— Neuwied, 243.
— non-arsenical, 275.
— patent, 250.
— pigments, compounded, 273.
— pigments, examination of, 442.
— Plessy's, 266.
— Rosenstiehl's, 270.
— sap, 410.
— Scheele's, 241.
— Schnitzer's, 266.
— silk, 275.
— Turkish, 265.
— verditer, 243.
— Vienna, 248.
Guignet's green, 264.
Gum Arabic, 420.
Guyard's violet, 278.
Gypsum, 40.

HACHETT'S brown, 280.
Hæmateïn, 406.
Hæmatoxylin, 406.
Humins, 283.
Hydrometer, 24.

INDIAN ink, 316.
— madder, 378.
— red, 185.
— yellow, 352.
Indigo, 390.
— carmine, 394.
Introduction, 1.
Iron brown, 280.
— compounds, 54.

Iron red, 182.
Ivory black, 290.

KUHLMANN's green, 294.

LAC, 363.
— dye, 363.
Lake, blue, 390, 397.
— Florentine, 361.
— fustic, 351.
— gamboge, 349.
— green, 409.
— quercitron, 351.
— weld, 349.
Lakes, 6, 9, 343.
— examination of, 445.
— madder, 375.
— red, 354.
— yellow, 348.
Lamp black, 307.
Lead acetate, 62.
— antimonate, 115.
— antimonite, 115.
— arsenite, 159.
— brown, 279.
— chloride, 64.
— chromate, 134.
— compounds, 60.
— iodide, 154.
— monoxide, 143.
— nitrate, 61.
— orange, 146.
— oxychloride, 113.
— red, 144.
— sulphate, 61, 112.
— sulphite, 114.
— tungstate, 128.
— white, 73, 74, 75, 77.
— white, hard and soft, 191.
Leaf green, 265.
Lichens, 380.
Lime, 39.
— blue, 228.
— green, 250.
Litharge, 143.
Lithophone, 119.
Litmus, 382.
Logwood, 398.
Lokao, 44.

MADDER, 370.
— carmine, 376.
— extract, 372.

Madder, Indian, 378.
— lakes, 375.
Magnesia, gold purple, 192.
— white, 130.
Magnesium carbonate, 41.
Manganese blue, 272.
— brown, 279.
— compounds, 59.
— green, 270.
— sulphate, 56.
— violet, 278.
— white, 130.
Manganous oxide, 272.
Mangit, 378.
Mars yellow, 155.
Massicot, 143.
Mercuric ammonium chloride, 175.
— chloride, 68.
— iodide, 176.
— nitrate, 68.
— sulphite, 163.
Mercurous chloride, 68.
— nitrate, 67.
Mercury compounds, 67.
— yellow, 158.
Metallic pigments, 326.
Metals, heavy, 51.
Mills, paint, 430.
— white lead, 86.
Mineral lake, 278.
Molybdenum blue, 239.
— compounds, 59.
Montpellier yellow, 148.
Mosaic gold, 160.

NAPLES yellow, 149.
Natural green, 275.
Neuberg blue, 227.
Neutral tint black, 318.
Neuwied green, 243.
Nickel compounds, 56.
— yellow, 157.
Non-arsenical green, 275.

OIL blue, 228.
Orange lead, 146.
— Mars, 155.
Orceïn, 300.
Orcinol, 380.
Orpiment, 159.

PAINT mills, 430.
Patent green, 250.

Payen's mountain blue, 228.
Permanent yellow, 152.
Pigments, artificial mineral, 6.
— commercial names of, 460.
— earth, 6.
Pine black, 312.
Pink, Dutch, 348.
Plessy's green, 266.
Poisonous pigments, 13.
Potassium alum, 45.
— bichromate, 35.
— bitartrate, 34.
— carbonate, 32.
— compounds, 32.
— ferricyanide, 35.
— ferrocyanide, 35.
— hydroxide, 33.
— nitrate, 34.
— sodium chromate, 35.
Precipitation, 7.
Preparation of pigments, 429.
Prepared gamboge, 350.
Prussian blue, 199.
— brown, 280.
Purple, alumina gold, 193.
— French, 381.
— magnesia gold, 192.
— of Cassius, 190.
— red, 189.
Purpurin, 373.
Purree, 352.
Pyrolusite brown, 280.

QUERCITRON lake, 351.
Quicklime, 39.

REALGAR, 159.
Red, carthamine, 366.
— chica, 379.
— chrome, 186.
— cobalt, 188.
— — magnesia, 188.
— enamels, 322.
— hæmatite, 180.
— Indian, 185.
— iron, 182.
— lakes, 354.
— lead, 144.
— Mars, 155.
— mineral pigments, 163.
— pigments, examination of, 440.
— purple, 189.

Red wood lakes, 384.
Rosenstiehl's green, 270.

SAFFLOWER, 366.
— carmine, 367.
Saffron, 352.
Sal ammoniac, 25.
Salt, common, 38.
Sandalwood, 388.
Sap colours, 416.
— green, 410.
Scheele's green, 241.
Schnitzer's green, 266.
Sepia, 414.
Shell gold, 326.
Shell silver, 327.
Siderin yellow, 156.
Silk green, 275.
Silver chromate, 189.
— compounds, 69.
— imitation, 328.
Smalts, 233.
Sodium chloride, 38.
— hydroxide, 37.
— salts, 37.
— thiosulphate, 38.
Soluble Prussian blue, 200.
Soot black, 313.
— pigments, 294.
Stannic chloride, 59.
Stannous chloride, 59.
Stove, drying, 459.
Sulphuretted hydrogen, 26.
Swedish green, 241.

TESSIÉ du Motay's blue, 239.
Thallium pigments, 159.
Thenard's blue, 230.
Tin compounds, 59.
— violet, 278.
— white, 130.
Tragacanth, 420.
Tungsten blue, 238.
— bronze pigments, 338.
— compounds, 59.
— yellow, 157.
Turkish green, 265.
Turnbull's blue, 203.
Turner's yellow, 149.
Turpeth mineral, 158.

ULTRAMARINE, 3, 204, 211.
— artificial, 206.

Ultramarine, cobalt, 230.
— natural, 205.
— violet, 219.
— yellow, 152.

VANADIUM compounds, 59.
Vegetable bronze pigments, 339.
Verdigris, 252.
— blue, 252.
— distilled, 255.
— German, 258.
Verditer, green, 243.
Vermilion, 3, 163, 166.
— antimony, 178.
— Chinese, 170.
— chrome, 186.
Vienna green, 248.
Vine black, 288.
Violet copper, 278.
— enamel, 324.
— Guyard's, 278.
— manganese, 278.
— mineral pigments, 276.
— Nuremberg, 278.
— tin, 278.
— ultramarine, 219.
Vitriol, blue, 65.
— green, 54.
— oil of, 27.

WASHING apparatus, 120.
Water, 16.
— colours, 419.
— — moist, 422.
— examination of, 20.
— hard, 17.
— iron in, 18.
— organic matter in, 18.
Weld lake, 349.
White, bismuth, 130.
— enamel, 116, 121, 320.
— lead, 73, 74, 75.

White lead mills, 861.
— magnesia, 131.
— manganese, 130.
— mineral pigments, 72.
— patent, 73.
— permanent, 116, 121.
— pigments, examination of, 437.
— tin, 130.
— tungsten, 128.
— zinc, 126.
Witherite, 42, 117.

YELLOW, antimony, 151.
— barium, 152.
— cadmium chrome, 153.
— calcium chrome, 151.
— Cassel, 148.
— chrome, 134.
— enamels, 322.
— English, 149.
— Indian, 352.
— lakes, 348.
— Mars, 155.
— Mercury, 158.
— mineral pigments, 133.
— Montpellier, 148.
— Naples, 149.
— nickel, 157.
— pale chrome, 139.
— pigments, examination of, 439.
— Siderin, 156.
— tungsten, 157.
— Turner's, 149.
— zinc chrome, 152.

ZINC chrome yellow, 152.
— oxide, 53.
— sulphate, 53.
— sulphide, 119.
— white, 126.
— — Griffith's, 128.

Lightning Source UK Ltd.
Milton Keynes UK
UKHW021936050420
361370UK00005B/9